T0214881

Power Systems

For further volumes:
http://www.springer.com/series/4622

Hyo J. Eom

Primary Theory
of Electromagnetics

 Springer

Hyo J. Eom
Department of Electrical Engineering
KAIST
Daejeon
Republic of Korea

ISSN 1612-1287   ISSN 1860-4676   (electronic)
ISBN 978-94-024-0032-8   ISBN 978-94-007-7143-7   (eBook)
DOI 10.1007/978-94-007-7143-7
Springer Dordrecht Heidelberg New York London

Printed on acid-free paper

Springer is part of Springer Science+Business Media (www.springer.com)

Preface

Electromagnetics is a fascinating branch of science dealing with electricity and magnetism. Historically, many renowned scientists have made important contributions to electromagnetics, which is a foundation of modern electrical engineering, information, and communications. Current wireless communications using mobile phones, for instance, epitomize the importance of electromagnetics. Electromagnetics covers many formulas and laws, which relate electric and magnetic fields to electric charges and currents. This text is intended for an introductory electromagnetic course for undergraduate students majoring in electrical engineering. The goal of this text is to lead students away from memorization, but toward a deeper understanding of formulas. To achieve this goal, many formulas commonly used for electromagnetic analysis are mathematically derived from a few fundamental empirical laws. To derive formulas mathematically is as much important as to apply them to practical, real problems. Basic knowledge in advanced calculus is required to derive formulas. Physical interpretations of formulas are intentionally de-emphasized for fear of unnecessary confusion. Each important formula is framed to indicate its significance.

We will present electromagnetics in chronological order from static fields to time-varying fields. Topics are divided into four parts: static electric fields, static magnetic fields, time-varying fields, and electromagnetic waves. Static electric fields and static magnetic fields are introduced from Coulomb's law and Ampère's law of force, respectively. Time-varying fields and waves are discussed in terms of Maxwell's equations, which include Faraday's law, Ampère's law, and two Gauss's laws. Chapter 1 introduces basic vector calculus, which is an indispensable tool for describing electromagnetic fields. Chapter 2 starts from charges and Coulomb's law to deal with static electric fields. Chapter 3 begins with steady currents and Ampère's law of force to discuss static magnetic fields. Chapter 4 discusses Faraday's law of induction. Chapter 5 presents time-varying fields based on Maxwell's equations. Chapter 6 presents uniform plane wave propagation. Chapter 7 presents transmission line theory. Chapter 8 presents waveguide and antenna fundamentals. The International System of Units (SI) is tacitly understood throughout the text. Symbols, notations, and units used in this text are summarized in Appendix A.

I wish to thank students at KAIST for their comments and suggestions on the manuscript. I also wish to thank Professors Y. H. Cho, J. K. Park, and Y. B. Park for reading the manuscript and providing valuable comments.

Daejeon, Korea, May 2013   Hyo J. Eom

Contents

Chapter 1
Vectors

1.1 Vector Operations

Quantities carrying direction and magnitude are defined as vectors. Vector examples include velocity, force, and acceleration. On the other hand, quantities carrying magnitude alone are defined as scalars. Time and mass are examples of scalars. Vectors in this text are denoted by symbols with overlines. A vector $\overline{A}$ is represented by

$$\overline{A} = \hat{A} A \tag{1.1}$$

where A is the magnitude of vector $\overline{A}$ and $\hat{A}$ is a direction-designating unit vector:

$$A = |\overline{A}| \tag{1.2}$$

$$\hat{A} = \frac{\overline{A}}{A} . \tag{1.3}$$

A vector $\overline{A}$ is graphically shown in Fig. 1.1, where the length of an arrow represents the magnitude of $\overline{A}$ and its arrowhead points to the direction of $\overline{A}$. Vectors $\overline{A}$ and $\overline{B}$ are equal if their directions and magnitudes are the same.

1.1.1 Vector Addition and Scalar Multiplication

The addition of vectors $\overline{A}$ and $\overline{B}$ is

$$\overline{C} = \overline{A} + \overline{B} . \tag{1.4}$$

A graphical representation of vector addition is shown in Fig. 1.2 . The multiplication of vector $\overline{A}$ by a positive real number c is $c\overline{A}$, which represents a vector whose

H. J. Eom, *Primary Theory of Electromagnetics*, Power Systems,
DOI: 10.1007/978-94-007-7143-7_1, © Springer Science+Business Media Dordrecht 2013

Fig. 1.1 Graphical representation of a vector $\overline{A}$

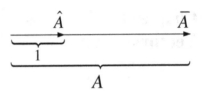

Fig. 1.2 The vector addition
$\overline{C} = \overline{A} + \overline{B}$, where $\overline{C}$ is the
diagonal of the parallelogram
made by $\overline{A}$ and $\overline{B}$

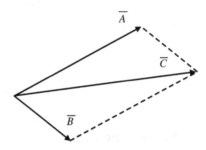

magnitude is c times as large as $\overline{A}$, while its direction remains unaltered. Furthermore,
$-\overline{B}$ is a vector whose magnitude is the same as $\overline{B}$ but its direction is opposite to $\overline{B}$.

1.1.2 Vector Multiplications

We define the scalar (or dot) product of two vectors $\overline{A}$ and $\overline{B}$ as

$$\overline{A} \cdot \overline{B} = AB \cos \theta \qquad (1.5)$$

scalar product

where $A = |\overline{A}|$, $B = |\overline{B}|$, and θ denotes a smaller angle between $\overline{A}$ and $\overline{B}$. A
graphical representation of the scalar product $\overline{A} \cdot \overline{B}$ is shown in Fig. 1.3. We define
the vector (or cross) product of vectors $\overline{A}$ and $\overline{B}$ as

$$\overline{A} \times \overline{B} = \hat{n} AB \sin \theta \qquad (1.6)$$

vector product

Fig. 1.3 The scalar product $C = \overline{A} \cdot \overline{B}$, where $C \geq 0$ if $0 \leq \theta \leq \pi/2$ and $C < 0$ if $\pi/2 < \theta \leq \pi$

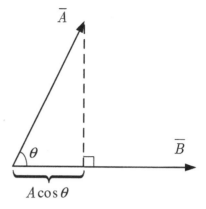

Fig. 1.4 The vector product $\overline{C} = \overline{A} \times \overline{B}$

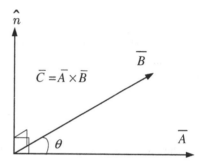

where θ is a smaller angle between $\overline{A}$ and $\overline{B}$ and $\hat{n}$ is a unit vector perpendicular to the plane that is made of vectors $\overline{A}$ and $\overline{B}$. The unit vector $\hat{n}$ advances in the same direction as a right-handed screw that rotates $\overline{A}$ through the angle θ to coincide with $\overline{B}$. A graphical representation of the vector product $\overline{A} \times \overline{B}$ is shown in Fig. 1.4.

1.2 Orthogonal Coordinates

Three commonly used orthogonal coordinate systems for electromagnetic field analysis are rectangular (x, y, z), cylindrical (ρ, ϕ, z), and spherical (r, θ, ϕ) coordinate systems. The rectangular coordinate system is most familiar to us but the cylindrical and spherical coordinate systems are also useful for the analysis of long circular cylinders and spheres. First, a rectangular coordinate representation is illustrated.

1.2.1 Rectangular (Cartesian) Coordinates

A vector $\overline{A}$ is written in terms of unit vectors $\hat{x}$, $\hat{y}$, and $\hat{z}$ as

$$\overline{A} = \hat{x} A_x + \hat{y} A_y + \hat{z} A_z \tag{1.7}$$

Fig. 1.5 Vector $\overline{A}$ in rectangular coordinates

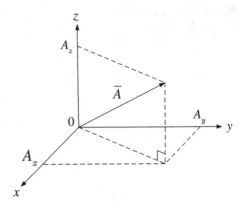

where A_x, A_y, and A_z are the components of $\overline{A}$. A graphical representation is shown in Fig. 1.5. The directions of unit vectors $\hat{x}$, $\hat{y}$, and $\hat{z}$ are designated by the axes of rectangular coordinates x, y, and z, respectively. By the Pythagorean theorem we write the magnitude of vector $\overline{A}$ as

$$|\overline{A}| = A = \left(A_x^2 + A_y^2 + A_z^2\right)^{1/2}. \tag{1.8}$$

The unit vectors $\hat{x}$, $\hat{y}$, and $\hat{z}$ are mutually orthogonal with the following cyclic properties:

$$\hat{x} \times \hat{y} = \hat{z} \tag{1.9}$$

$$\hat{y} \times \hat{z} = \hat{x} \tag{1.10}$$

$$\hat{z} \times \hat{x} = \hat{y}. \tag{1.11}$$

When $\overline{B} = \hat{x} B_x + \hat{y} B_y + \hat{z} B_z$, the scalar product $\overline{A} \cdot \overline{B}$ produces

$$\overline{A} \cdot \overline{B} = (\hat{x} A_x + \hat{y} A_y + \hat{z} A_z) \cdot (\hat{x} B_x + \hat{y} B_y + \hat{z} B_z)$$

$$= A_x B_x \underbrace{(\hat{x} \cdot \hat{x})}_{1} + A_x B_y \underbrace{(\hat{x} \cdot \hat{y})}_{0} + A_x B_z \underbrace{(\hat{x} \cdot \hat{z})}_{0} + \cdots$$

$$= A_x B_x + A_y B_y + A_z B_z. \tag{1.12}$$

Similarly the vector product $\overline{A} \times \overline{B}$ is

Fig. 1.6 Differential length $d\bar{r}$, where $\bar{r}$ is a position vector

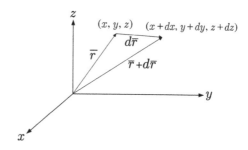

$$\overline{A} \times \overline{B} = (\hat{x}A_x + \hat{y}A_y + \hat{z}A_z) \times (\hat{x}B_x + \hat{y}B_y + \hat{z}B_z)$$

$$= A_xB_x \underbrace{(\hat{x} \times \hat{x})}_{0} + A_xB_y \underbrace{(\hat{x} \times \hat{y})}_{\hat{z}} + A_xB_z \underbrace{(\hat{x} \times \hat{z})}_{-\hat{y}} + \cdots$$

$$= \hat{x}(A_yB_z - A_zB_y) + \hat{y}(A_zB_x - A_xB_z)$$

$$+ \hat{z}(A_xB_y - A_yB_x). \tag{1.13}$$

The vector product is compactly rewritten as

$$\overline{A} \times \overline{B} = \begin{vmatrix} \hat{x} & \hat{y} & \hat{z} \\ A_x & A_y & A_z \\ B_x & B_y & B_z \end{vmatrix}. \tag{1.14}$$

A position vector is used to designate a specific point in space. We will introduce a position vector using rectangular coordinates. Figure 1.6 shows a position vector $\bar{r} = \hat{x}x + \hat{y}y + \hat{z}z$ pointing from the origin to the position (x, y, z). When the position moves over a differential distance from (x, y, z) to $(x + dx, y + dy, z + dz)$, the differential length vector is given by

$$d\bar{r} = \hat{x}\,dx + \hat{y}\,dy + \hat{z}\,dz. \tag{1.15}$$

Figure 1.7 illustrates an infinitesimal rectangular box (rectangular parallelepiped) of dx, dy, and dz. The differential areas, normal to the directions $\hat{x}$, $\hat{y}$, and $\hat{z}$, are $ds_x = dy\,dz$, $ds_y = dz\,dx$, and $ds_z = dx\,dy$, respectively; the differential volume is expressed as $dv = dx\,dy\,dz$.

1.2.2 Cylindrical Coordinates

Orthogonal curvilinear coordinates: It is sometimes necessary to use orthogonal curvilinear coordinates different from the rectangular coordinates. First consider

Fig. 1.7 Differential volume
dv in rectangular coordinates

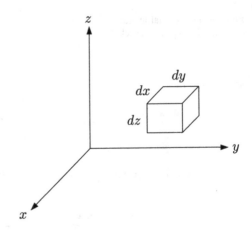

a general case of orthogonal curvilinear coordinates described by (u_1, u_2, u_3). The
position vector $\bar{r} = \hat{x}x + \hat{y}y + \hat{z}z$ in space can be given by the curvilinear coordinates
(u_1, u_2, u_3) where

$$x = x(u_1, u_2, u_3) \tag{1.16}$$

$$y = y(u_1, u_2, u_3) \tag{1.17}$$

$$z = z(u_1, u_2, u_3). \tag{1.18}$$

The u_1 coordinate curve can be generated by a position vector when u_1 varies with
u_2 and u_3 being held constant. Similarly the u_2 and u_3 coordinate curves can be
generated. The derivative $\dfrac{\partial \bar{r}}{\partial u_i}$ is a tangent vector to the u_i coordinate curve. We
define the unit vector $\hat{u}_i$ as

$$\hat{u}_i = \frac{\dfrac{\partial \bar{r}}{\partial u_i}}{\left| \dfrac{\partial \bar{r}}{\partial u_i} \right|}. \tag{1.19}$$

When $\hat{u}_1$, $\hat{u}_2$, and $\hat{u}_3$ are perpendicular to each other, the curvilinear coordinate
systems are referred to as orthogonal.

The first example of the orthogonal curvilinear coordinates is the cylindrical coordinates (ρ, ϕ, z), as shown in Fig. 1.8. Note that ρ is a radial distance from the z-axis
and ϕ is referred to as an azimuth angle on the x-y plane. The relations between
(x, y, z) and (ρ, ϕ, z) are

Fig. 1.8 Position vector $\bar{r}$ in cylindrical (ρ, ϕ, z) coordinates

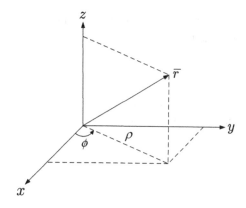

$$x = \rho \cos \phi \tag{1.20}$$

$$y = \rho \sin \phi \tag{1.21}$$

$$z = z \tag{1.22}$$

$$\rho = \sqrt{x^2 + y^2} \tag{1.23}$$

$$\phi = \tan^{-1} \frac{y}{x} \tag{1.24}$$

$$z = z. \tag{1.25}$$

The position vector can be represented in terms of (ρ, ϕ, z) as

$$\bar{r} = \hat{x}x + \hat{y}y + \hat{z}z$$

$$= \hat{x}\rho \cos \phi + \hat{y}\rho \sin \phi + \hat{z}z \tag{1.26}$$

where their ranges are given by

$$0 \leq \rho < \infty \tag{1.27}$$

$$0 \leq \phi < 2\pi \tag{1.28}$$

$$-\infty < z < \infty. \tag{1.29}$$

Any vector $\overline{A}$ can be expressed in terms of the cylindrical coordinate unit vectors $\hat{\rho}$, $\hat{\phi}$, and $\hat{z}$ as

$$\overline{A} = \hat{\rho} A_\rho + \hat{\phi} A_\phi + \hat{z} A_z. \tag{1.30}$$

The unit vectors $\hat{\rho}$ and $\hat{\phi}$ can be given in terms of $\hat{x}$ and $\hat{y}$. Assuming $u_1 = \rho$, $u_2 = \phi$, and $u_3 = z$, we obtain

$$\frac{\partial \overline{r}}{\partial \rho} = \hat{x} \cos \phi + \hat{y} \sin \phi \tag{1.31}$$

$$\frac{\partial \overline{r}}{\partial \phi} = -\hat{x} \rho \sin \phi + \hat{y} \rho \cos \phi \tag{1.32}$$

$$\frac{\partial \overline{r}}{\partial z} = \hat{z}. \tag{1.33}$$

Hence

$$\hat{\rho} = \frac{\dfrac{\partial \overline{r}}{\partial \rho}}{\left| \dfrac{\partial \overline{r}}{\partial \rho} \right|} = \hat{x} \cos \phi + \hat{y} \sin \phi \tag{1.34}$$

$$\hat{\phi} = \frac{\dfrac{\partial \overline{r}}{\partial \phi}}{\left| \dfrac{\partial \overline{r}}{\partial \phi} \right|} = -\hat{x} \sin \phi + \hat{y} \cos \phi. \tag{1.35}$$

We note that the directions of $\hat{\rho}$ and $\hat{\phi}$ vary as ϕ changes. The unit vectors satisfy the following cyclic properties:

$$\hat{\rho} \times \hat{\phi} = \hat{z} \tag{1.36}$$

$$\hat{\phi} \times \hat{z} = \hat{\rho} \tag{1.37}$$

$$\hat{z} \times \hat{\rho} = \hat{\phi}. \tag{1.38}$$

Fig. 1.9 Differential volume
dv in cylindrical coordinates

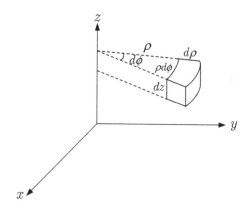

The magnitude of vector $\overline{A}$ is

$$A = \left(A_\rho^2 + A_\phi^2 + A_z^2 \right)^{1/2}. \tag{1.39}$$

The differential length vector $d\overline{r}$ is written as

$$d\overline{r} = \underbrace{\frac{\partial \overline{r}}{\partial \rho}}_{\hat{\rho}} d\rho + \underbrace{\frac{\partial \overline{r}}{\partial \phi}}_{\hat{\phi}\rho} d\phi + \underbrace{\frac{\partial \overline{r}}{\partial z}}_{\hat{z}} dz$$

$$= \hat{\rho}\,d\rho + \hat{\phi}\rho\,d\phi + \hat{z}\,dz. \tag{1.40}$$

Figure 1.9 shows the differential volume constructed from the differential lengths, $d\rho$, $\rho\,d\phi$, and dz:

$$dv = \rho\,d\rho\,d\phi\,dz. \tag{1.41}$$

1.2.3 Spherical Coordinates

The spherical coordinates (r, θ, ϕ) are shown in Fig. 1.10, where their relations to (x, y, z) are

$$x = r \sin\theta \cos\phi \tag{1.42}$$

$$y = r \sin\theta \sin\phi \tag{1.43}$$

$$z = r \cos\theta \tag{1.44}$$

Fig. 1.10 Position vector $\bar{r}$ in
spherical (r, θ, ϕ) coordinates

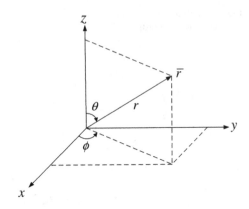

$$r = \sqrt{x^2 + y^2 + z^2} \tag{1.45}$$

$$\theta = \tan^{-1} \frac{\sqrt{x^2 + y^2}}{z} \tag{1.46}$$

$$\phi = \tan^{-1} \frac{y}{x}. \tag{1.47}$$

Note that r is the distance from the origin, θ is called the polar angle, and ϕ is the azimuth angle. The polar angle is measured from the positive z-axis. The azimuth angle in spherical coordinates is identical with the azimuth angle in cylindrical coordinates. The position vector can be represented in terms of (r, θ, ϕ) as

$$\bar{r} = \hat{x}x + \hat{y}y + \hat{z}z$$

$$= \hat{x}r \sin \theta \cos \phi + \hat{y}r \sin \theta \sin \phi + \hat{z}r \cos \theta \tag{1.48}$$

where their ranges are given by

$$0 \leq r < \infty \tag{1.49}$$

$$0 \leq \theta \leq \pi \tag{1.50}$$

$$0 \leq \phi < 2\pi. \tag{1.51}$$

A vector $\overline{A}$ is represented in terms of the spherical coordinates (r, θ, ϕ) as

$$\overline{A} = \hat{r}A_r + \hat{\theta}A_\theta + \hat{\phi}A_\phi. \tag{1.52}$$

We can write $\hat{r}$, $\hat{\theta}$, and $\hat{\phi}$ in terms of $\hat{x}$, $\hat{y}$, and $\hat{z}$. Substituting (1.42) through (1.44) into (1.19), we obtain

$$\hat{r} = \hat{x}\sin\theta\cos\phi + \hat{y}\sin\theta\sin\phi + \hat{z}\cos\theta \tag{1.53}$$

$$\hat{\theta} = \hat{x}\cos\theta\cos\phi + \hat{y}\cos\theta\sin\phi - \hat{z}\sin\theta \tag{1.54}$$

$$\hat{\phi} = -\hat{x}\sin\phi + \hat{y}\cos\phi. \tag{1.55}$$

Also note that the directions of $\hat{r}$, $\hat{\theta}$, and $\hat{\phi}$ vary as the coordinates θ and ϕ change. The unit vectors $\hat{r}$, $\hat{\theta}$, and $\hat{\phi}$ satisfy the following cyclic properties:

$$\hat{r} \times \hat{\theta} = \hat{\phi} \tag{1.56}$$

$$\hat{\theta} \times \hat{\phi} = \hat{r} \tag{1.57}$$

$$\hat{\phi} \times \hat{r} = \hat{\theta}. \tag{1.58}$$

The magnitude of vector $\overline{A}$ is given by

$$A = \left(A_r^2 + A_\theta^2 + A_\phi^2\right)^{1/2}. \tag{1.59}$$

The differential length vector $d\overline{r}$ is given by

$$d\overline{r} = \underbrace{\frac{\partial \overline{r}}{\partial r}}_{\hat{r}}\, dr + \underbrace{\frac{\partial \overline{r}}{\partial \theta}}_{\hat{\theta}r}\, d\theta + \underbrace{\frac{\partial \overline{r}}{\partial \phi}}_{\hat{\phi}r\sin\theta}\, d\phi$$

$$= \hat{r}\,dr + \hat{\theta}r\,d\theta + \hat{\phi}r\sin\theta\,d\phi. \tag{1.60}$$

Figure 1.11 shows the differential volume dv constructed from the differential lengths dr, $r\,d\theta$, and $r\sin\theta\,d\phi$. It is

$$dv = r^2\sin\theta\,dr\,d\theta\,d\phi. \tag{1.61}$$

Fig. 1.11 Differential volume
dv in spherical coordinates

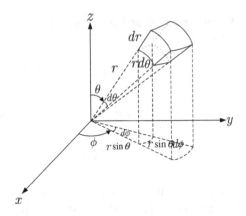

1.3 Operator Del, Divergence Theorem, and Stokes's Theorem

The gradient, divergence, curl, and Laplacian are differential expressions developed for vector calculus. All these expressions permit us to describe electromagnetic fields compactly and efficiently. We will introduce these expressions using rectangular coordinates without providing their physical interpretations. In addition, we will present the divergence theorem and Stokes's theorem, which are two indispensable tools widely used throughout this text.

1.3.1 Gradient, Divergence, Curl, and Laplacian

We first introduce the differential vector operator ∇, called *del*, as

$$\nabla = \hat{x}\frac{\partial}{\partial x} + \hat{y}\frac{\partial}{\partial y} + \hat{z}\frac{\partial}{\partial z}. \tag{1.62}$$

The gradient of a scalar function f is defined as

$$\nabla f = \hat{x}\frac{\partial f}{\partial x} + \hat{y}\frac{\partial f}{\partial y} + \hat{z}\frac{\partial f}{\partial z}. \tag{1.63}$$

The divergence of a vector function $\overline{A} \ (= \hat{x}A_x + \hat{y}A_y + \hat{z}A_z)$ is defined as

$$\nabla \cdot \overline{A} = \frac{\partial A_x}{\partial x} + \frac{\partial A_y}{\partial y} + \frac{\partial A_z}{\partial z}. \tag{1.64}$$

The curl of a vector function $\overline{A}$ is defined as

$$\nabla \times \overline{A} = \begin{vmatrix} \hat{x} & \hat{y} & \hat{z} \\ \dfrac{\partial}{\partial x} & \dfrac{\partial}{\partial y} & \dfrac{\partial}{\partial z} \\ A_x & A_y & A_z \end{vmatrix}. \tag{1.65}$$

The Laplacian of a scalar function f is defined as

$$\nabla^2 f = \nabla \cdot \nabla f = \frac{\partial^2 f}{\partial x^2} + \frac{\partial^2 f}{\partial y^2} + \frac{\partial^2 f}{\partial z^2}. \tag{1.66}$$

The vector Laplacian in rectangular coordinates is shown to be

$$\nabla^2 \overline{A} = \hat{x} \nabla^2 A_x + \hat{y} \nabla^2 A_y + \hat{z} \nabla^2 A_z. \tag{1.67}$$

Gradients, divergences, curls, and Laplacians in cylindrical and spherical coordinates can be derived by using a chain rule for a function of several variables. Such derivations may be tedious but straightforward. Final expressions for gradients, divergences, curls, and Laplacians in cylindrical and spherical coordinates are summarized in Appendix C. These differential expressions are used for the problems dealing with long cylindrical and spherical structures.

1.3.2 Divergence Theorem

The divergence theorem deals with the divergence of a vector over a volume surrounded by a closed surface. Figure 1.12 shows a vector $\overline{A}$ passing through a volume

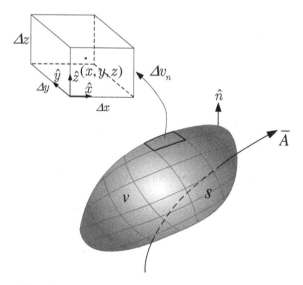

Fig. 1.12 A vector $\overline{A}$ passing through a volume v bounded by a closed surface s

v bounded by a closed surface s. The vector $\overline{A}$ and its first partial derivatives are continuous over the region containing v. We subdivide v into infinitesimal rectangular parallelepipeds $\Delta v_n = \Delta x \, \Delta y \, \Delta z$ centered at (x, y, z). The divergence of $\overline{A}$ over Δv_n yields

$$
\int_{\Delta v_n} \nabla \cdot \overline{A} \, dv
$$
$$
= \int_{z - \frac{\Delta z}{2}}^{z + \frac{\Delta z}{2}} \int_{y - \frac{\Delta y}{2}}^{y + \frac{\Delta y}{2}} \int_{x - \frac{\Delta x}{2}}^{x + \frac{\Delta x}{2}} \left(\frac{\partial A_x}{\partial x} + \frac{\partial A_y}{\partial y} + \frac{\partial A_z}{\partial z} \right) dx \, dy \, dz. \tag{1.68}
$$

Note

$$
\int_{x - \frac{\Delta x}{2}}^{x + \frac{\Delta x}{2}} \frac{\partial A_x}{\partial x} \, dx = A_x(x + \frac{\Delta x}{2}, y, z) - A_x(x - \frac{\Delta x}{2}, y, z). \tag{1.69}
$$

Therefore (1.68) becomes

$$
\int_{\Delta v_n} \nabla \cdot \overline{A} \, dv
$$
$$
= \left[A_x(x + \frac{\Delta x}{2}, y, z) - A_x(x - \frac{\Delta x}{2}, y, z) \right] \Delta y \, \Delta z
$$
$$
+ \left[A_y(x, y + \frac{\Delta y}{2}, z) - A_y(x, y - \frac{\Delta y}{2}, z) \right] \Delta x \, \Delta z
$$
$$
+ \left[A_z(x, y, z + \frac{\Delta z}{2}) - A_z(x, y, z - \frac{\Delta z}{2}) \right] \Delta x \, \Delta y. \tag{1.70}
$$

The right-hand side of (1.70) can be compactly rewritten as $\oint_{\Delta s_n} \overline{A} \cdot d\overline{s}_n$, where the symbol $\oint_{\Delta s_n}$ denotes integration over a closed surface Δs_n surrounding a volume Δv_n and

$$
\Delta \overline{s}_n = \begin{cases} \pm \hat{x} \, \Delta y \, \Delta z \text{ for planes normal to } \hat{x} \text{ at } (x \pm \frac{\Delta x}{2}, y, z) \\ \pm \hat{y} \, \Delta z \, \Delta x \text{ for planes normal to } \hat{y} \text{ at } (x, y \pm \frac{\Delta y}{2}, z) \\ \pm \hat{z} \, \Delta x \, \Delta y \text{ for planes normal to } \hat{z} \text{ at } (x, y, z \pm \frac{\Delta z}{2}). \end{cases} \tag{1.71}
$$

Here the direction of $\Delta \overline{s}_n$ has been chosen to point outward from Δv_n. Hence

$$
\int_{\Delta v_n} \nabla \cdot \overline{A} \, dv = \oint_{\Delta s_n} \overline{A} \cdot d\overline{s}_n. \tag{1.72}
$$

Fig. 1.13 Infinitesimal
surfaces Δs_n

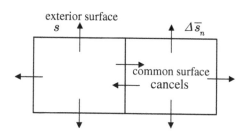

The summation of $\nabla \cdot \overline{A}$ over v is

$$\underbrace{\sum_{n=1}^{\infty} \int_{\Delta v_n} \nabla \cdot \overline{A} \, dv}_{\displaystyle \int_{v} \nabla \cdot \overline{A} \, dv} = \sum_{n=1}^{\infty} \oint_{\Delta s_n} \overline{A} \cdot d\overline{s}_n. \qquad (1.73)$$

Figure 1.13 shows infinitesimal surfaces Δs_n belonging to either the exterior surface s or the common surface between adjacent rectangular volumes. When infinitesimal surfaces Δs_n belong to the common surfaces having two opposite directions, $\int_{\Delta s_n} \overline{A} \cdot d\overline{s}_n$ is canceled. Therefore, the right-hand side of (1.73) becomes $\oint_{s} \overline{A} \cdot d\overline{s}$, which is the integration over the exterior surface s alone. We rewrite (1.73) as

$$\int_{v} \nabla \cdot \overline{A} \, dv = \oint_{s} \overline{A} \cdot d\overline{s} \qquad (1.74)$$

divergence theorem

where $d\overline{s} = \hat{n} ds$ and $\hat{n}$ points to the direction outwardly normal to the surface ds and away from the volume v, as shown in Fig. 1.12. The divergence theorem states that the outward flux $\left(\oint_{s} \overline{A} \cdot d\overline{s} \right)$ through the closed surface s equals the volume integration of $\nabla \cdot \overline{A}$. The divergence theorem will be extensively utilized later in this text to develop formulations and laws that are fundamental in electromagnetic theory. In the following example, the divergence theorem is utilized to derive an important formula relating the divergence to the Dirac delta functions. The derivation is tricky but worthy of note.

Fig. 1.14 Spherical coordinates (R, ϑ, φ)

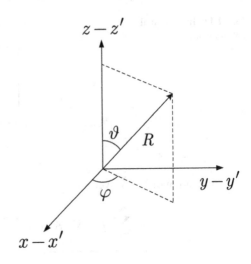

Example 1.1   Divergence theorem application.

Evaluate $f(\overline{R}) = \nabla \cdot \left(\dfrac{\hat{R}}{R^2} \right)$, where $\hat{R} = \dfrac{\overline{R}}{|\overline{R}|}$ and

$$\overline{R} = \overline{r} - \overline{r}' = \hat{x}(x - x') + \hat{y}(y - y') + \hat{z}(z - z'). \qquad (1.75)$$

Solution: It is convenient to introduce the spherical coordinates (R, ϑ, φ) centered at $(x = x', y = y', z = z')$, as shown in Fig. 1.14 where

$$R = \sqrt{(x - x')^2 + (y - y')^2 + (z - z')^2}. \qquad (1.76)$$

For any vector $\overline{F}$ in (R, ϑ, φ) coordinates, its divergence is given by

$$\nabla \cdot \overline{F} = \frac{1}{R^2} \frac{\partial (R^2 F_R)}{\partial R} + \frac{1}{R \sin \vartheta} \frac{\partial (\sin \vartheta F_\vartheta)}{\partial \vartheta} + \frac{1}{R \sin \vartheta} \frac{\partial F_\varphi}{\partial \varphi}. \qquad (1.77)$$

Let us evaluate $f(\overline{R})$ when $R \neq 0$ and when $R \to 0$ separately.

1. When $R \neq 0$,

$$f(\overline{R}) = \frac{1}{R^2} \frac{\partial (R^2 R^{-2})}{\partial R} = 0. \qquad (1.78)$$

2. When $R \to 0$, $f(\overline{R})$ approaches an indeterminate form (0/0), which requires a careful treatment. To investigate its behavior near $R = 0$, we consider a small sphere v with a radius R, as shown in Fig. 1.15. We integrate $f(\overline{R})$ over v and apply the divergence theorem to obtain

Fig. 1.15 A small sphere of a volume v surrounded by a surface s

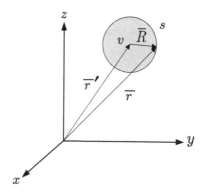

$$\int_v f(\overline{R})\, dv = \int_v \nabla \cdot \left(\frac{\hat{R}}{R^2} \right) dv = \oint_s \left(\frac{\hat{R}}{R^2} \right) \cdot d\overline{s}$$

$$= \frac{\hat{R}}{R^2} \cdot 4\pi R^2 \hat{R}$$

$$= 4\pi. \tag{1.79}$$

Expressions (1.78) and (1.79) lead to

$$\nabla \cdot \left(\frac{\hat{R}}{R^2} \right) = 4\pi \delta(\overline{R}) \tag{1.80}$$

$$\delta(\overline{R}) = \delta(x - x')\, \delta(y - y')\, \delta(z - z'). \tag{1.81}$$

Expression (1.80) is a useful relation, which enables us to handle point charges and currents in the form of Dirac delta functions. Some more discussion on the Dirac delta functions is given in Appendix D.

1.3.3 Stokes's Theorem

Stokes's theorem deals with the curl of a vector over an open surface. Figure 1.16 illustrates a vector $\overline{A}$ passing through an open surface s bounded by a closed contour l. The vector $\overline{A}$ and its first partial derivatives are continuous over the region containing s. The surface s is subdivided into small differentialrectangular areas Δs_n.

Fig. 1.16 A vector $\overline{A}$ passing through an open surface s

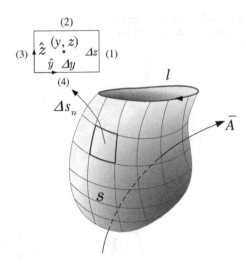

The curl of $\overline{A}$ is

$$
\begin{aligned}
\nabla \times \overline{A} \\
= \hat{x}\left(\frac{\partial A_z}{\partial y} - \frac{\partial A_y}{\partial z}\right) + \hat{y}\left(\frac{\partial A_x}{\partial z} - \frac{\partial A_z}{\partial x}\right) + \hat{z}\left(\frac{\partial A_y}{\partial x} - \frac{\partial A_x}{\partial y}\right).
\end{aligned}
\tag{1.82}
$$

First, we write the x-component of (1.82) as

$$
\begin{aligned}
(\nabla \times \overline{A}) \cdot \hat{x} = {} & \frac{A_z(x, y + \frac{\Delta y}{2}, z) - A_z(x, y - \frac{\Delta y}{2}, z)}{\Delta y} \\
& - \frac{A_y(x, y, z + \frac{\Delta z}{2}) - A_y(x, y, z - \frac{\Delta z}{2})}{\Delta z}
\end{aligned}
\tag{1.83}
$$

where (y, z) is the center of a differential rectangular area $\Delta s_n = \Delta y \, \Delta z$, as shown in Fig. 1.16. We define the differential length vector $\Delta \overline{r}_n$ encircling Δs_n as

$$
\Delta \overline{r}_n =
\begin{cases}
+\hat{z}\Delta z & \text{for line (1)} \\[1em]
-\hat{z}\Delta z & \text{for line (3)} \\[1em]
-\hat{y}\Delta y & \text{for line (2)} \\[1em]
+\hat{y}\Delta y & \text{for line (4).}
\end{cases}
\tag{1.84}
$$

Fig. 1.17 Differential length
vector $\Delta \bar{r}_n$

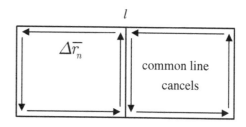

Note that the directions of $\Delta \bar{r}_n$ and $\hat{x} \Delta s_n$ follow a right-hand rule. According to the right-hand rule, when the right-hand fingers adhere to the $\Delta \bar{r}_n$, the thumb designates the $\hat{x}$-direction. Therefore (1.83) is rewritten as

$$
\left(\nabla \times \overline{A} \right) \cdot \hat{x} = \frac{\oint_{\Delta r_n} \overline{A} \cdot d\bar{r}_n}{\Delta s_n} \tag{1.85}
$$

where $\oint_{\Delta r_n}$ denotes a line integral along the closed path Δr_n. We obtain similar expressions for the y- and z-components of (1.82). The summation of $\nabla \times \overline{A}$ over s is

$$
\underbrace{\sum_{n=1}^{\infty} \left(\nabla \times \overline{A} \right) \cdot \hat{u}_n \Delta s_n}_{\int_s \left(\nabla \times \overline{A} \right) \cdot d\bar{s}} = \sum_{n=1}^{\infty} \oint_{\Delta r_n} \overline{A} \cdot d\bar{r}_n \tag{1.86}
$$

where $\hat{u}_n = \hat{x}$, $\hat{y}$, or $\hat{z}$. The right-hand side of (1.86) amounts to a line integral along the closed contour l since line integrals are canceled along the common lines between two adjacent rectangular areas Δs_n (see Fig. 1.17). Therefore (1.86) becomes

$$
\int_s \left(\nabla \times \overline{A} \right) \cdot d\bar{s} = \oint_l \overline{A} \cdot d\bar{r} \tag{1.87}
$$

Stokes's theorem

where the directions of $d\bar{r}$ and $d\bar{s}$ follow the right-hand rule (see Fig. 1.18). Stokes's theorem states that the circulation $\left(\oint_l \overline{A} \cdot d\bar{r} \right)$ through the closed path l equals the surface integration of $\nabla \times \overline{A}$ over the surface s bounded by l. Stokes's theorem will be utilized later in the text to formulate important electromagnetic laws.

Fig. 1.18 Directions of $d\bar{r}$
and $d\bar{s}$

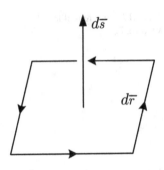

1.4 Problems for Chapter 1

1. Express the unit vector $\hat{x}$ in terms of the spherical coordinate unit vectors $(\hat{r}, \hat{\theta}, \hat{\phi})$.
2. Prove (1.60).
3. Derive the expression of ∇f using the cylindrical coordinates (ρ, ϕ, z).
   Hint: Use a chain rule $\dfrac{\partial f}{\partial x} = \dfrac{\partial f}{\partial \rho}\dfrac{\partial \rho}{\partial x} + \dfrac{\partial f}{\partial \phi}\dfrac{\partial \phi}{\partial x}$.
4. Derive the expression of $\nabla^2 f$ using the spherical coordinates (r, θ, ϕ).
5. Show $\nabla \times (\nabla f) = 0$.
6. Show $\nabla \cdot (\nabla \times \overline{F}) = 0$.
7. Show $\nabla \cdot (\overline{A} \times \overline{B}) = \overline{B} \cdot (\nabla \times \overline{A}) - \overline{A} \cdot (\nabla \times \overline{B})$.
8. Show $\nabla \times \nabla \times \overline{A} = -\nabla^2 \overline{A} + \nabla (\nabla \cdot \overline{A})$.
9. Evaluate $\oint_l \overline{r} \cdot d\overline{r}$ where $\overline{r} = \hat{x}x + \hat{y}y + \hat{z}z$. Here, the symbol $\oint_l$ denotes a line integral along any closed path l.
10. Evaluate $\nabla\left(\dfrac{1}{R}\right)$ where $R = \sqrt{(x - x')^2 + (y - y')^2 + (z - z')^2}$.

Chapter 2
Electrostatics

2.1 Fundamentals of Electric Fields

2.1.1 Coulomb's Law and Electric Fields

It has been known that electric charges exert forces on each other according to
Coulomb's law (Fig. 2.1). Coulomb's law states the amount of forces between two
point charges. Figure 2.2 illustrates two static point charges, q_1 and q_2, separated
in unbounded free space by a distance R. Ideally spatial sizes of point charges are
negligible compared to R. The charges q_1 and q_2 are located at positions given by
the two position vectors as

$$\bar{r}' = \hat{x}x' + \hat{y}y' + \hat{z}z' \tag{2.1}$$
$$\bar{r} = \hat{x}x + \hat{y}y + \hat{z}z \tag{2.2}$$

where

$$\overline{R} = \bar{r} - \bar{r}'$$
$$= \hat{x}(x - x') + \hat{y}(y - y') + \hat{z}(z - z'). \tag{2.3}$$

The force between charges is inversely proportional to R^2 and is directly proportional to $q_1 q_2$. The types of charges are either positive or negative. The same types
($q_1 q_2 > 0$) repel but the opposite ($q_1 q_2 < 0$) attract. In summary q_1 exerts a force
$\overline{F}$ on q_2 as

$$\overline{F} = \hat{R}\frac{q_1 q_2}{4\pi\epsilon_0 R^2} \tag{2.4}$$

Coulomb's law

H. J. Eom, *Primary Theory of Electromagnetics*, Power Systems,
DOI: 10.1007/978-94-007-7143-7_2, © Springer Science+Business Media Dordrecht 2013

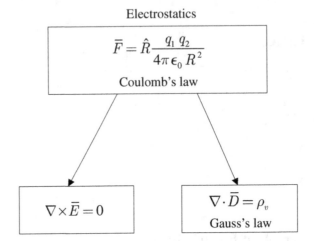

Electrostatics

$$\overline{F} = \hat{R} \frac{q_1 q_2}{4\pi \epsilon_0 R^2}$$

Coulomb's law

$$\nabla \times \overline{E} = 0$$

$$\nabla \cdot \overline{D} = \rho_v$$

Gauss's law

Fig. 2.1 Fundamental equations for static electric fields

Fig. 2.2 Two point charges q_1 and q_2 in free space. The distance between them is $R = |\overline{R}|$. The force $\overline{F}$ is assumed repulsive for illustration

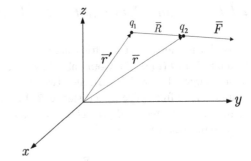

where the unit vector $\hat{R}$ is given by

$$\hat{R} = \frac{\overline{R}}{|\overline{R}|}. \tag{2.5}$$

Here ϵ_0 is the permittivity of free space (vacuum) as

$$\epsilon_0 \approx 8.854 \times 10^{-12} \approx \frac{10^{-9}}{36\pi} \; (\text{F/m}). \tag{2.6}$$

Next we consider a point charge distribution Q that is located near another point charge q_{test}, as shown in Fig. 2.3. Note that q_{test} at position P experiences an electric force $\overline{F}$ due to the distribution of Q. According to the superposition principle the force $\overline{F}$ due to Q is a vector sum of forces due to q_i ($i = 1, \cdots, N$). A helpful way of dealing with this situation is the idea of field. We define the electric field $\overline{E}$ at position P as

Fig. 2.3 A test charge q_{test} placed near a charge distribution Q

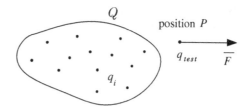

$$E = \frac{\overline{F}}{q_{test}}. \tag{2.7}$$

definition of electric field

Note that $\overline{E}$ is a function of Q without reference to q_{test}.

1. When a single point charge q is located at $\overline{r}'$, the electric field $\overline{E}$ at $\overline{r}$ due to q is

$$\overline{E} = \hat{R}\frac{q}{4\pi\epsilon_0 R^2} \tag{2.8}$$

where $\overline{R} = \overline{r} - \overline{r}'$. The direction of $\overline{E}$ from a positive point charge q points radially outward $\hat{R}$ from q.

2. Based on (2.8), the electric field for a continuous charge distribution can be obtained. Figure 2.4 shows a continuous charge volume v' with a volume charge density ρ_v' located at $\overline{r}'$. The volume charge density is defined as

$$\rho_v' = \lim_{\Delta v' \to 0} \frac{\Delta Q}{\Delta v'} \tag{2.9}$$

where ΔQ is a differential charge contained within a differential volume $\Delta v'$ located at $\overline{r}'$. In other words, the volume charge density is regarded as a charge per unit volume. The electric field $\overline{E}$ at $\overline{r}$ is

$$\overline{E} = \int_{v'} \hat{R}\frac{\rho_v'}{4\pi\epsilon_0 R^2} \, dv' \tag{2.10}$$

where the symbol $\int_{v'}$ denotes a volume integration over v'.

3. Consider a special case when a continuous charge distribution shrinks to a point charge q. The charge density for a point charge q located at $\overline{r}' = \overline{r}''$ is given by

$$\rho_v' = q\,\delta\left(\overline{r}' - \overline{r}''\right) \tag{2.11}$$

Fig. 2.4 Continuous charge
distribution ρ'_v in free space

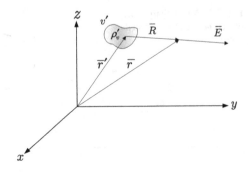

where $\delta \left(\bar{r}' - \bar{r}'' \right)$ is the Dirac delta function. In rectangular coordinates, this is
given as

$$\delta \left(\bar{r}' - \bar{r}'' \right) = \delta(x' - x'') \, \delta(y' - y'') \, \delta(z' - z''). \qquad (2.12)$$

Substitution of (2.11) into (2.10) yields

$$\bar{E} = \int_{v'} \hat{R} \frac{q \delta \left(\bar{r}' - \bar{r}'' \right)}{4 \pi \epsilon_0 R^2} \, dv' = \hat{R} \frac{q}{4 \pi \epsilon_0 R^2} \qquad (2.13)$$

where $\bar{R} = \bar{r} - \bar{r}''$. Note that (2.10) is reduced to a point charge case of (2.8).

2.1.2 Electric Scalar Potential

The electric scalar potential is a useful tool for evaluating electric fields. The electric
scalar potential is also known as the electrostatic potential or the electric potential for
short. Although the electric potential is customarily defined as the potential energy
per unit charge, the electric potential can be also introduced directly from charges.
To this end, we will consider the charge distribution ρ'_v whose electric field is given
by (2.10). Since

$$\nabla \left(\frac{1}{R} \right) = \nabla \left[\frac{1}{\sqrt{(x - x')^2 + (y - y')^2 + (z - z')^2}} \right]$$

$$= -\frac{\hat{x}(x - x') + \hat{y}(y - y') + \hat{z}(z - z')}{\left[(x - x')^2 + (y - y')^2 + (z - z')^2 \right]^{1.5}}$$

$$= -\frac{\hat{R}}{R^2} \qquad (2.14)$$

we rewrite (2.10) as

$$\overline{E} = -\int_{v'} \nabla\left(\frac{1}{R}\right) \frac{\rho'_v}{4\pi\epsilon_0} \, dv' \tag{2.15}$$

where ρ'_v has been tacitly assumed to be a function of $\overline{r}'$. The operator ∇ deals with differentiation with respect to (x, y, z) whereas the operator $\int_{v'}$ denotes integration with respect to (x', y', z'). Therefore we can interchange these two operators as

$$\overline{E} = -\frac{1}{4\pi\epsilon_0} \nabla \int_{v'} \frac{\rho'_v}{R} \, dv'. \tag{2.16}$$

At this point, we introduce the electric scalar potential in terms of charges as

$$V = \frac{1}{4\pi\epsilon_0} \int_{v'} \frac{\rho'_v}{R} \, dv'. \tag{2.17}$$

electric scalar potential in terms of charges

We note that ρ'_v is a function of (x', y', z') and V is a function of (x, y, z). The electric scalar potential at (x, y, z) is obtained by summing up the contributions from charges over (x', y', z'). The electric potential V becomes zero at infinity $(R \to \infty)$ as long as the extent of charge distribution remains finite. The electric field is rewritten in terms of V as

$$\overline{E} = -\nabla V. \tag{2.18}$$

electric field in terms of electric scalar potential

The electric scalar potential should be regarded as an intermediate step for evaluating the electric field, which is our ultimate goal. Thanks to the vector identity $\nabla \times (\nabla V) \equiv 0$, we get

$$\nabla \times \overline{E} = 0. \tag{2.19}$$

From this we will introduce the concept of potential difference. Consider two points A and B in a vector field $\overline{E}$, as shown in Fig. 2.5. Connecting A and B with two arbitrary paths Γ_1 and Γ_2, and integrating $\nabla \times \overline{E}$ over a surface s enclosed by these two paths, we obtain

Fig. 2.5 Γ_1 and Γ_2 connecting points A and B in the electric field $\overline{E}$

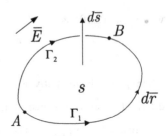

$$\int_s \nabla \times \overline{E} \cdot d\overline{s} = 0. \qquad (2.20)$$

We apply Stokes's theorem to (2.20) to obtain

$$\oint_{\Gamma_1-\Gamma_2} \overline{E} \cdot d\overline{r} = 0 \qquad (2.21)$$

where the symbol $\oint_{\Gamma_1-\Gamma_2}$ denotes a line integration around a closed path in the counterclockwise direction. We further note that (2.21) yields

$$\int_{\Gamma_1} \overline{E} \cdot d\overline{r} - \int_{\Gamma_2} \overline{E} \cdot d\overline{r} = 0 \Longrightarrow$$

$$\int_{\Gamma_1} \overline{E} \cdot d\overline{r} = \int_{\Gamma_2} \overline{E} \cdot d\overline{r}. \qquad (2.22)$$

The line integrals $\left(\int_{\Gamma_1} \text{ and } \int_{\Gamma_2} \right)$ are independent of the choice of paths since the paths were initially arbitrarily chosen. Next we consider a differential of V as

$$dV = \underbrace{\frac{\partial V}{\partial x}dx + \frac{\partial V}{\partial y}dy + \frac{\partial V}{\partial z}dz}_{\nabla V \cdot d\overline{r}} = -\overline{E} \cdot d\overline{r}. \qquad (2.23)$$

Integrating dV along the path Γ_1 connecting two points A and B in Fig. 2.5, we obtain

$$\underbrace{\int_{V_A}^{V_B} dV}_{V_B - V_A} = -\int_A^B \overline{E} \cdot d\overline{r} \qquad (2.24)$$

where V_A and V_B are electric scalar potentials at A and B, respectively. Letting $V_B - V_A = V_{BA}$, we introduce the electric scalar potential difference

Fig. 2.6 Electric scalar potential V due to a point charge q

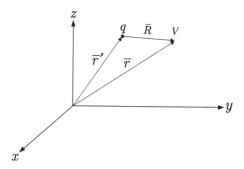

$$V_{BA} = -\int_A^B \overline{E} \cdot d\overline{r}. \qquad (2.25)$$

electric scalar potential difference

Note that V_{BA} represents the work required to move a unit charge from A to B in the field of $\overline{E}$. If a reference point A recedes to infinity of zero potential ($V_A = 0$), the electric scalar potential at position B is given by

$$V_B = -\int_\infty^B \overline{E} \cdot d\overline{r}. \qquad (2.26)$$

This is a representation of the electric scalar potential in terms of the electric field. The electric scalar potential difference (electric potential difference or simply potential difference) is referred to as a voltage. It is seen that a voltage around any closed path must be zero: this is called Kirchhoff's voltage law in circuit theory.

Example 2.1 Electric scalar potential due to a point charge.
Find the electric scalar potential V at $\overline{r}$ due to a point charge q at $\overline{r}'$, as shown in Fig. 2.6.

Solution: The electric scalar potential at the position $\overline{r}$ is

$$V = \frac{1}{4\pi\epsilon_0} \int_{v'} \frac{\rho_v'}{R} \, dv'. \qquad (2.27)$$

Since $\rho_v' = q\,\delta\left(\overline{r}' - \overline{r}''\right)$, the electric scalar potential is

$$V = \frac{q}{4\pi\epsilon_0 R} \qquad (2.28)$$

where $R = |\overline{r} - \overline{r}'|$.

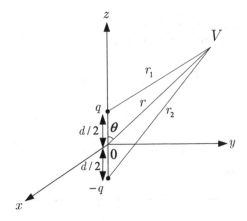

Fig. 2.7 Two point charges
$\pm q$ with a distance d apart:
electric dipole

Example 2.2   Electric field due to an electric dipole.

Figure 2.7 shows an electric dipole, a pair of point charges, $+q$ and $-q$, separated
by a distance d. Evaluate the electric field when $r \gg d$.

Solution: Two charges $\pm q$ are regarded as independent sources that produce electric fields. Therefore, the superposition principle applies, and the potential at r is the
sum of two potential contributions from $+q$ and $-q$:

$$V = \frac{q}{4\pi\epsilon_0}\left(\frac{1}{r_1} - \frac{1}{r_2}\right). \tag{2.29}$$

Since $r \gg d$, r_1 and r_2 are approximately given by

$$r_1 \approx r - \frac{d}{2}\cos\theta \tag{2.30}$$

$$r_2 \approx r + \frac{d}{2}\cos\theta. \tag{2.31}$$

Substituting (2.30) and (2.31) into (2.29), we obtain the electric scalar potential at
far distance

$$V \approx \frac{q}{4\pi\epsilon_0}\left(\frac{d\cos\theta}{\underbrace{r^2 - \frac{d^2}{4}\cos^2\theta}_{\ll r^2}}\right) \approx \frac{qd\cos\theta}{4\pi\epsilon_0 r^2} \tag{2.32}$$

which shows that the potential decreases as the square of the distance from the dipole.
The electric field is written in terms of the spherical coordinates (r and θ) as

Fig. 2.8 An electric dipole moment $\overline{p}$

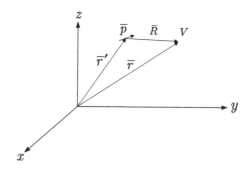

$$\overline{E} = -\nabla V$$

$$= -\left(\hat{r}\frac{\partial V}{\partial r} + \hat{\theta}\frac{1}{r}\frac{\partial V}{\partial \theta}\right)$$

$$= \frac{qd}{4\pi\epsilon_0 r^3}\left(\hat{r}2\cos\theta + \hat{\theta}\sin\theta\right). \tag{2.33}$$

It is convenient to rewrite (2.32) in compact form by introducing the electric dipole moment $\overline{p} = q\overline{d}$, where $\overline{d}$ is a distant vector (magnitude: d, direction: pointing from $-q$ to q). The electric dipole moment $\overline{p}$ in Fig. 2.8 illustrates an electric dipole located at $\overline{r}'$. The electric scalar potential at $\overline{r}$ due to $\overline{p}$ is written as

$$V = \frac{\overline{p}\cdot\hat{R}}{4\pi\epsilon_0 R^2} \tag{2.34}$$

where $\overline{R} = \overline{r} - \overline{r}'$ and $\hat{R}$ is its unit vector. Some of molecules composed of atoms are regarded as dipoles. Molecules exhibiting dipole moments such as water are called polar molecules.

2.1.3 Gauss's Law

Gauss's law is one of fundamental equations relating electrostatic fields to charges. We will derive Gauss's law by considering the divergence of electric field:

$$\nabla \cdot \overline{E} = \int_{v'} \frac{\rho'_v}{4\pi\epsilon_0} \underbrace{\nabla \cdot \left(\frac{\hat{R}}{R^2}\right)}_{4\pi\delta(\overline{R})} dv'$$

$$= \frac{\rho'_v}{\epsilon_0}\bigg|_{\overline{r}'\to\overline{r}}$$

$$= \frac{\rho_v}{\epsilon_0} \tag{2.35}$$

Fig. 2.9 Volume charge Q
occupying a volume v with a
Gaussian surface s

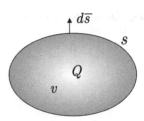

where the volume charge densities ρ'_v and ρ_v are tacitly assumed to be functions of $\bar{r}'$ and $\bar{r}$, respectively. In other words, $\rho'_v = \rho_v(\bar{r}')$ and $\rho_v = \rho_v(\bar{r})$. Refer to Example 1.1 for the relation $\nabla \cdot \left(\dfrac{\hat{R}}{R^2} \right) = 4\pi\delta(\bar{R})$, where $\delta(\bar{R}) = \delta(x - x')\,\delta(y - y')\,\delta(z - z')$ is the three-dimensional Dirac delta function. Thus we have

$$\epsilon_0 \nabla \cdot \overline{E} = \rho_v. \tag{2.36}$$

Gauss's law in differential form

It is possible to convert Gauss's law in differential form into Gauss's law in integral form that is also widely used. We consider a volume charge Q (volume charge density ρ_v) occupying a volume v, as shown in Fig. 2.9. Integrating Gauss's law over v, we obtain

$$\epsilon_0 \int_v \nabla \cdot \overline{E}\, dv = \underbrace{\int_v \rho_v\, dv}_{Q}. \tag{2.37}$$

Although the integration has been performed over the volume occupied by ρ_v, we may include any exterior region devoid of charges ($\rho_v = 0$), if necessary. Applying the divergence theorem to the Gaussian surface s enclosing the charge Q, we obtain

$$\epsilon_0 \oint_s \overline{E} \cdot d\overline{s} = Q. \tag{2.38}$$

Gauss's law in integral form

Gauss's law states that the outward electric flux $\left(\epsilon_0 \oint_s \overline{E} \cdot d\overline{s} \right)$ passing through a closed surface s equals the charge enclosed by the surface. Gauss's law is effective in computing electric fields especially when the distribution of charges is symmetric, as shown in the next example.

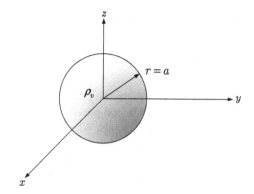

Fig. 2.10 A spherical charge with a uniform charge density ρ_v of radius a

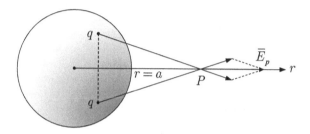

Fig. 2.11 Two identical charges q with respect to r

Example 2.3 Electric field due to a charged sphere.

Figure 2.10 illustrates a sphere of radius a uniformly charged with a volume charge density ρ_v. Evaluate the electric field by using Gauss's law.

Solution: First consider two identical charges q symmetric with respect to r in Fig. 2.11. The resultant electric field $\overline{E}_p$ at position P points to the r-direction. Next consider a spherical charge in Fig. 2.10. Due to spherical symmetry, the electric field has a component E_r in the radial direction.

1. When $r > a$, we construct a Gaussian spherical surface surrounding ρ_v. Applying Gauss's law with a radius $r > a$, we obtain

$$\epsilon_0 E_r 4\pi r^2 = \frac{4}{3}\pi a^3 \rho_v \implies E_r = \frac{\rho_v a^3}{3\epsilon_0 r^2}. \tag{2.39}$$

2. When $r \le a$, we construct a Gaussian spherical surface with a radius $r \le a$. The electric field for $r \le a$ is

$$\epsilon_0 E_r 4\pi r^2 = \frac{4}{3}\pi r^3 \rho_v \implies E_r = \frac{\rho_v}{3\epsilon_0} r. \tag{2.40}$$

2.2 Dielectrics and Boundary Conditions

2.2.1 Dielectrics

Good dielectrics refer to materials whose electric charges are so tightly bound that no electric current can flow. Dielectrics are composed of positively charged nuclei surrounded by negatively charged electron clouds. When electric fields are applied to dielectrics, molecules within dielectrics experience forces according to Coulomb's law. Coulomb forces displace each positively charged nucleus and negatively charged electron cloud differently to generate realigned dipoles. This scenario is illustrated in Fig. 2.12. Consider a differential volume dv' that possesses electric dipole moment. If the dipole moment per unit volume is given by a polarization vector $\overline{P}'\ (= \overline{P}(\overline{r}'))$, the differential dipole moment possessed by dv' is $\overline{P}'\,dv'$. The differential electric potential at $\overline{r}$ due to dv' at $\overline{r}'$ (refer to (2.34)) is

$$dV = \frac{\overline{P}' \cdot \hat{R}}{4\pi\epsilon_0 R^2}\,dv' \tag{2.41}$$

where $R = |\overline{r} - \overline{r}'|$. The total potential due to v' is

$$V = \int_{v'} \frac{\overline{P}' \cdot \hat{R}}{4\pi\epsilon_0 R^2}\,dv'. \tag{2.42}$$

Since $\nabla'\left(\dfrac{1}{R}\right) = \dfrac{\hat{R}}{R^2}$, where $\nabla' = \hat{x}\dfrac{\partial}{\partial x'} + \hat{y}\dfrac{\partial}{\partial y'} + \hat{z}\dfrac{\partial}{\partial z'}$, the potential is rewritten as

$$V = \frac{1}{4\pi\epsilon_0} \int_{v'} \overline{P}' \cdot \nabla'\left(\frac{1}{R}\right) dv'. \tag{2.43}$$

Furthermore we note the vector identity

$$\nabla' \cdot \left(\frac{\overline{P}'}{R}\right) = \overline{P}' \cdot \nabla'\left(\frac{1}{R}\right) + \frac{\nabla' \cdot \overline{P}'}{R}. \tag{2.44}$$

Fig. 2.12 Molecules under an externally applied field $\overline{E}$

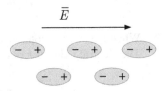

Hence

$$V = \frac{1}{4\pi\epsilon_0} \left[\int_{v'} \nabla' \cdot \left(\frac{\overline{P}'}{R} \right) dv' + \int_{v'} \frac{-\nabla' \cdot \overline{P}'}{R} dv' \right]$$

$$= \frac{1}{4\pi\epsilon_0} \left[\oint_{s'} \frac{1}{R} \underbrace{\overline{P}' \cdot \hat{n}'}_{\rho'_{ps}} ds' + \int_{v'} \frac{1}{R} \underbrace{\left(-\nabla' \cdot \overline{P}' \right)}_{\rho'_{pv}} dv' \right] \qquad (2.45)$$

where the divergence theorem has been used. We are now in a position to interpret the terms ρ'_{ps} and ρ'_{pv}. Since the electric scalar potential due to a volume charge density ρ'_v is

$$V = \int_{v'} \frac{\rho'_v}{4\pi\epsilon_0 R} dv' \qquad (2.46)$$

the terms ρ'_{ps} and ρ'_{pv} in (2.45) are regarded as a bound surface charge density and a bound volume charge density, respectively. Bound charges associated with ρ'_{ps} and ρ'_{pv} are not allowed to move freely as opposed to free charges. The electric scalar potential due to a polarized dielectric can be obtained in terms of ρ'_{ps} and ρ'_{pv}.

Gauss's law in dielectrics: For the sake of generality, we consider a dielectric volume that contains free charges (free volume charge density ρ_v) in addition to bound charges (bound volume charge density ρ_{pv}). Note that Gauss's law relates the electric field $\overline{E}$ within a dielectric volume to the volume charge density $\left(\rho_v + \rho_{pv} \right)$ as

$$\nabla \cdot \left(\epsilon_0 \overline{E} \right) = \rho_v + \rho_{pv}. \qquad (2.47)$$

Substituting $\nabla \cdot \left(-\overline{P} \right) = \rho_{pv}$ into (2.47) gives

$$\nabla \cdot \left(\epsilon_0 \overline{E} + \overline{P} \right) = \rho_v. \qquad (2.48)$$

It has been tacitly understood that $\overline{E}$, ρ_v, ρ_{pv}, and $\overline{P}$ in (2.47) and (2.48) are all functions of unprimed coordinates (x, y, z). Next we introduce the electric displacement (or electric flux density)

$$\overline{D} = \epsilon_0 \overline{E} + \overline{P} \qquad (2.49)$$

to rewrite (2.48) as

$$\nabla \cdot \overline{D} = \rho_v. \qquad (2.50)$$

Gauss's law in dielectrics

This is Gauss's law for dielectric materials whose free charge density is ρ_v. The polarization vector $\overline{P}$ in most dielectrics is proportional to $\overline{E}$ as

$$\overline{P} = \chi_e \epsilon_0 \overline{E} \tag{2.51}$$

where χ_e is the electric susceptibility of a dielectric. The value of χ_e for a linear, homogeneous dielectric medium is constant. Therefore the relation between $\overline{D}$ and $\overline{E}$ is

$$\overline{D} = \epsilon_0 \underbrace{(1 + \chi_e)}_{\epsilon_r} \overline{E} = \epsilon \overline{E} \tag{2.52}$$

where $\epsilon = \epsilon_0 \epsilon_r$ is the permittivity of a dielectric, and ϵ_r is the relative permittivity (dielectric constant). Integrating Gauss's law over the volume

$$\int_v \nabla \cdot \overline{D}\, dv = \underbrace{\int_v \rho_v\, dv}_{Q} \tag{2.53}$$

and applying the divergence theorem, we obtain

$$\oint_s \overline{D} \cdot d\overline{s} = Q. \tag{2.54}$$

Gauss's law in dielectrics in integral form

Now we are in a position to reinterpret Gauss's law for dielectrics. Consider the electric flux density $\overline{D}$ in a dielectric medium enclosed by a surface s. Gauss's law in dielectrics indicates that the electric flux $\left(\oint_s \overline{D} \cdot d\overline{s} \right)$ passing outwardly through s is the same as the free charge Q enclosed by s.

2.2.2 Boundary Conditions

Boundary conditions refer to field behavior between different dielectrics. The boundary conditions are needed to obtain unique solutions to the boundary-value problems. The boundary conditions can be determined from the equations $\oint_l \overline{E} \cdot d\overline{r} = 0$ and $\oint_s \overline{D} \cdot d\overline{s} = Q$. Figure 2.13 shows a boundary separating dielectric medium 1 from

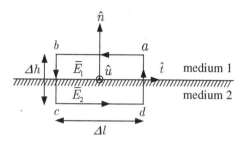

Fig. 2.13 Tangential boundary condition between dielectric medium 1 and medium 2

medium 2. Choosing the closed path l as $abcda$, we get

$$\oint_{abcda} \overline{E} \cdot d\overline{r} = 0. \tag{2.55}$$

When $\Delta h \to 0$, the integrations along the paths bc and da vanish. When $\overline{E}_1$ and $\overline{E}_2$ denote the electric fields in medium 1 and medium 2, respectively, (2.55) is rewritten as

$$\int_{ab} \overline{E}_1 \cdot d\overline{r} + \int_{cd} \overline{E}_2 \cdot d\overline{r} = 0 \implies -\underbrace{\overline{E}_1 \cdot \hat{t}}_{E_{1t}} \Delta l + \underbrace{\overline{E}_2 \cdot \hat{t}}_{E_{2t}} \Delta l = 0. \tag{2.56}$$

Therefore

$$E_{1t} = E_{2t}. \tag{2.57}$$

Since $\hat{t}$ can be any arbitrary tangential unit vector, (2.57) implies that the tangential electric field components should be continuous across the boundary between different media. Often, it is desirable to rewrite (2.57) in terms of $\hat{n}$, which is a unit vector normal to the boundary surface. Since $(\hat{n} \times \overline{E}_1)$ is another tangential vector, (2.57) can be written as

$$\hat{n} \times \left(\overline{E}_1 - \overline{E}_2\right) = 0. \tag{2.58}$$

tangential components of $\overline{E}$

Figure 2.14 shows an infinitesimal pill box with a base area Δs that is tangential to the boundary. The total charge within the pill box is ΔQ. Applying Gauss's law to the pill box, we obtain

$$\oint_s \overline{D} \cdot d\overline{s} = \Delta Q \tag{2.59}$$

which is rewritten in terms of top, bottom, and side faces as

Fig. 2.14 Normal boundary
condition between dielectric
medium 1 and medium 2

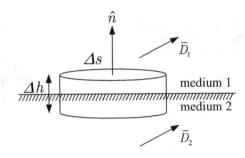

$$\int_{\text{top}} \overline{D} \cdot d\overline{s} + \int_{\text{bottom}} \overline{D} \cdot d\overline{s} + \int_{\text{side}} \overline{D} \cdot d\overline{s} = \Delta Q. \tag{2.60}$$

When $\Delta h \rightarrow 0$, then $\displaystyle\int_{\text{side}} \overline{D} \cdot d\overline{s} \rightarrow 0$ and (2.60) gives

$$\overline{D}_1 \cdot \hat{n} \Delta s - \overline{D}_2 \cdot \hat{n} \Delta s = \Delta Q. \tag{2.61}$$

Introducing the surface charge density

$$\rho_s = \lim_{\Delta s \to 0} \frac{\Delta Q}{\Delta s} \tag{2.62}$$

we obtain

$$\hat{n} \cdot \left(\overline{D}_1 - \overline{D}_2 \right) = \rho_s \, . \tag{2.63}$$

normal components of $\overline{D}$

The scalar product $\hat{n} \cdot \overline{D}_1$ takes a normal component of $\overline{D}_1$. Let D_{1n} and D_{2n} denote the normal components of $\overline{D}_1$ and $\overline{D}_2$, respectively, then the boundary condition is rewritten as

$$D_{1n} - D_{2n} = \rho_s. \tag{2.64}$$

The normal boundary condition requires the normal component of $\overline{D}$ to be discontinuous by a surface charge density ρ_s that is present at the boundary. Of course, if $\rho_s = 0$, the normal component of $\overline{D}$ is continuous across the boundary.

Fig. 2.15 Conductors in static electric fields

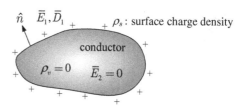

2.2.3 Conductors

Conductors are very special materials composed of free charges. Consider a conductor in a static electric field $\overline{E}_1$, as shown in Fig. 2.15. Free charges in the conductor move freely until the electric field $\overline{E}_2$ within the conductor becomes zero. In other words, the conductor interior achieves a zero-field state by redistributing free charges over the conductor surface in the form of surface charges. This means that free charges simply cannot exist inside the conductor and the volume charge density ρ_v inside the conductor must be zero; its rigorous proof is given in Example 5.1. We summarize the conductor characteristics as follows:

$$\overline{E}_2 = 0 \qquad (2.65)$$
$$\rho_v = 0. \qquad (2.66)$$

inside conductors

1. Since $\overline{E}_2 = 0$ inside conductors, the electric field tangential to conducting surfaces is also zero. The electrostatic potential along the conducting surface is, therefore, constant.
2. The normal component of $\overline{D}_1$ on the conductor surface is equal to the conductor surface charge density ρ_s. Therefore, the boundary conditions between a dielectric and a conductor are summarized as

$$\hat{n} \times \overline{E}_1 = 0 \qquad (2.67)$$
$$\hat{n} \cdot \overline{D}_1 = \rho_s \qquad (2.68)$$

boundary conditions for conductors

where $\hat{n}$ denotes a unit vector outwardly normal to the conductor surface.

Fig. 2.16 Infinitely long, charged, conducting cylinder of a radius a

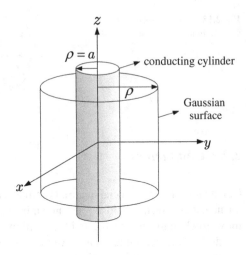

Example 2.4 Electric field due to a conducting cylinder.

In Fig. 2.16 an infinitely long conducting solid cylinder with a radius a carries a charge Q per unit length. The charge Q is uniformly distributed on the conductor surface at $\rho = a$. Determine the electric field.

Solution: We use the cylindrical coordinates (ρ, ϕ, z) for analysis. Since field distribution is symmetric with respect to the azimuth angle ϕ, it is convenient to use Gauss's law to determine the electric field. The electric field within the conductor is zero. When $\rho > a$, the electric field has a ρ-component dependent on ρ only due to geometrical symmetry. Choosing a circular cylindrical Gaussian surface with a radius $\rho > a$, we write Gauss's law $\left(\epsilon_0 \oint_s \overline{E} \cdot d\overline{s} = Q \right)$ as

$$\epsilon_0 \int_0^1 E_\rho 2\pi \rho \, dz = Q \implies E_\rho = \frac{Q}{2\pi \epsilon_0 \rho}. \tag{2.69}$$

Here the contributions from the top and bottom Gaussian surfaces are zero since $\overline{E}$ and $d\overline{s}$ are perpendicular to each other. Therefore the electric field is

$$E_\rho = \begin{cases} 0 & \text{for } \rho < a \\ \dfrac{Q}{2\pi \epsilon_0 \rho} & \text{for } \rho > a. \end{cases} \tag{2.70}$$

Note that the electric field is discontinuous at the conductor surface $\rho = a$.

2.3 Capacitance

A capacitor, sometimes called a condenser, is one of major components used in electronic circuits. For instance, very small capacitors serve as memory for the binary codes in the dynamic random access memory (DRAM) of digital computers. Capacitors connected to a time-varying voltage source store and release electric charges. If the time variation of a voltage source is sufficiently slow, a static electric field assumption should be valid for capacitor analysis. Figure 2.17 shows a DC voltage source V that supplies charges to two separate conductors. The conductors collect charges $\pm Q$ and an electric field is set up between the conductors. Note that the potential V is proportional to the charge Q. Here we introduce a constant C (capacitance) as

$$C = \frac{Q}{V}. \tag{2.71}$$

definition of capacitance

The charge-collecting device, a capacitor, is characterized by its capacitance C. Note that the capacitance is a function of capacitor geometries and medium permittivity.

Example 2.5 Capacitance between two conducting plates.
Figure 2.18 shows conducting parallel plates of area s and spacing d. The area s is large enough to ignore fringing effects. Determine the capacitance when the space between the plates is filled with a dielectric medium of permittivity ϵ.

Solution: Charges $\pm Q$ are uniformly distributed on the interior faces ($z = 0$ and $z = d$) and a uniform electric field $\overline{E}$ exists in the $-z$-direction. Selecting a Gaussian surface of rectangular parallelepiped to surround the upper plate, we write Gauss's law as

$$\oint_s \overline{D} \cdot d\overline{s} = Q \implies \epsilon E s = Q. \tag{2.72}$$

The potential difference and the capacitance are given by

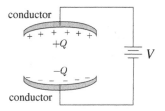

Fig. 2.17 Capacitor composed of two conducting plates

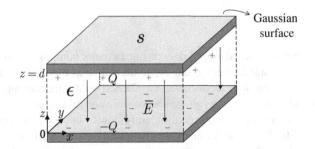

Fig. 2.18 Capacitance between two parallel conducting plates

$$V = -\int \overline{E} \cdot d\overline{r} = Ed \qquad (2.73)$$

$$C = \frac{Q}{V} = \frac{\epsilon s}{d}. \qquad (2.74)$$

Example 2.6 Capacitance of a coaxial cable.

Figure 2.19 shows a long coaxial cable composed of two conducting concentric cylinders of length l with radii $\rho = a$ and $\rho = b$. The outer conducting cylinder is infinitesimally thin. The space between the two concentric cylinders is filled with a dielectric material of permittivity ϵ. Determine the capacitance neglecting the fringe effect.

Fig. 2.19 Coaxial cable consisting of two conducting concentric cylinders $(l \gg b)$

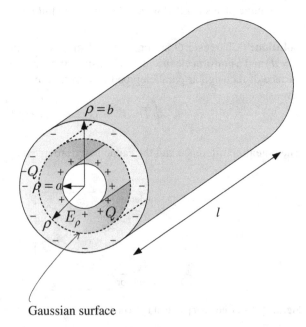

Solution: The electric potential difference between the inner and outer conducting cylinders is V and the total charge on the inner conducting cylinder is Q. When the inner conducting cylinder is surrounded by a cylindrical Gaussian surface with a radius $a < \rho < b$ and a length l, Gauss's law yields

$$\oint_S \underbrace{\overline{D}}_{\epsilon \overline{E}} \cdot d\bar{s} = Q \Longrightarrow E_\rho = \frac{Q}{2\pi\epsilon l\rho}. \tag{2.75}$$

Since

$$V = -\int \overline{E} \cdot d\bar{r} = -\int_b^a E_\rho\, d\rho = \frac{Q}{2\pi\epsilon l} \ln\left(\frac{b}{a}\right) \tag{2.76}$$

the capacitance is

$$C = \frac{Q}{V} = \frac{2\pi\epsilon l}{\ln\left(\dfrac{b}{a}\right)}. \tag{2.77}$$

2.4 Electrostatic Energy

It has been known that energy can be stored in electrostatic fields. Coulomb's law states that distributed charges exert forces on each other. In order to overcome these forces and to assemble charges, work is required and is stored in electric fields as electrostatic energy. Let us evaluate the amount of electrostatic energy stored in electric fields to assemble free charges. First consider a point charge in Fig. 2.20. A point charge q is brought from infinity to a position $\bar{r}$ against a Coulomb force

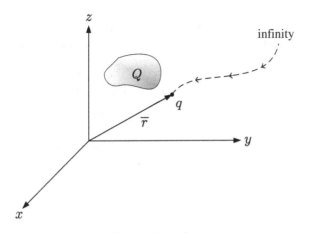

Fig. 2.20 A point charge q in the vicinity of a charge Q

Fig. 2.21 Distribution
of point charges q_m
$(m = 1, \cdots, N)$

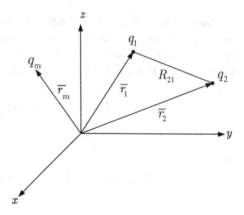

produced by a charge Q. A force $-\overline{F}$ must be applied to overcome the Coulomb
force $\overline{F} = q\overline{E}$. The work required to move the charge q against the Coulomb force is

$$W = \int_{\infty}^{\overline{r}} -\overline{F} \cdot d\overline{r} = -\int_{\infty}^{\overline{r}} q\overline{E} \cdot d\overline{r} = qV \qquad (2.78)$$

where V is an electric scalar potential at $\overline{r}$ with reference to infinity. We next consider
point charges in Fig. 2.21. Initially, the space is assumed devoid of charges. When
the first charge q_1 is brought to the final position $\overline{r}_1$ from infinity, no work is required
since no other charges were initially present. When the second charge q_2 is brought
to the final position $\overline{r}_2$, the work required to overcome the force due to q_1 is

$$W_2 = q_2 \underbrace{\frac{q_1}{4\pi\epsilon_0 R_{21}}}_{V_{21}} \qquad (2.79)$$

where $R_{21} = |\overline{r}_2 - \overline{r}_1|$ and V_{21} is the electric potential at $\overline{r}_2$ due to q_1. When the
third charge q_3 is brought to the final position $\overline{r}_3$, the required work is

$$W_3 = q_3 \left(\underbrace{\frac{q_1}{4\pi\epsilon_0 R_{31}}}_{V_{31}} + \underbrace{\frac{q_2}{4\pi\epsilon_0 R_{32}}}_{V_{32}} \right) \qquad (2.80)$$

where $R_{31} = |\overline{r}_3 - \overline{r}_1|$ and $R_{32} = |\overline{r}_3 - \overline{r}_2|$. Therefore, the total work required to
assemble the charges q_m $(m = 1, \cdots, N)$ is

$$W_e = q_2 V_{21} +$$
$$q_3 V_{31} + q_3 V_{32} +$$
$$q_4 V_{41} + q_4 V_{42} + q_4 V_{43} +$$
$$\cdots . \tag{2.81}$$

Since $q_m V_{mn} = q_n V_{nm}$, we rewrite the energy as

$$W_e = q_1 V_{12} +$$
$$q_1 V_{13} + q_2 V_{23} +$$
$$q_1 V_{14} + q_2 V_{24} + q_3 V_{34} +$$
$$\cdots . \tag{2.82}$$

Adding (2.81) to (2.82) gives

$$2W_e = q_1 (0 + V_{12} + V_{13} + V_{14} + \cdots) +$$
$$q_2 (V_{21} + 0 + V_{23} + V_{24} + \cdots) +$$
$$q_3 (V_{31} + V_{32} + 0 + V_{34} + \cdots) +$$
$$q_4 (V_{41} + V_{42} + V_{43} + 0 + \cdots) +$$
$$\cdots \tag{2.83}$$

which is rewritten as

$$W_e = \frac{1}{2} \sum_{m=1}^{N} q_m \underbrace{\sum_{n=1,n \neq m}^{N} V_{mn}}_{V_m} \tag{2.84}$$

where the notation V_m represents an electrostatic potential at position $\bar{r}_m$ due to all charges excluding q_m. Expression (2.84) gives the electrostatic energy required to assemble point charges q_m ($m = 1, \cdots , N$).

Next we will generalize (2.84) to the case of continuous charge distribution. Figure 2.22 illustrates an electrostatic system of free charge whose volume charge density is ρ_v in a volume v. In view of (2.84) the electrostatic energy stored within v is written as

$$W_e = \frac{1}{2} \int_v \rho_v V \, dv \tag{2.85}$$

where V represents the potential at $\bar{r}$ due to total charges that occupy v. It would be possible to extend v to an unbounded space by including an additional space devoid of charges ($\rho_v = 0$). Note that V is a potential due to continuous charge distribution whereas V_m in (2.84) is a potential due to discrete charge distribution. Is V compatible with V_m, which is the potential due to the charges except for q_m? The answer is yes for the following reason: in the case of V, the potential due to $\rho_v \, dv$ at

Fig. 2.22 Continuous volume
charge with a volume charge
density ρ_v

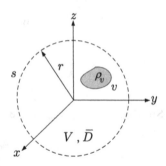

$\bar{r}$ is infinitesimal, and thus, is ignored. Therefore, V and V_m are surely compatible
with each other.

Energy in terms of fields in a dielectric medium: It should be noted that (2.85)
represents the work required to assemble free charges in free space. Next we will
derive the work required to assemble the free charges of volume charge density ρ_v
in a dielectric medium. Initially the dielectric medium is devoid of the free charges
and the free charges will be brought from infinity into the dielectric medium one by
one. In the following discussion, a symbol δ will be used to denote a small increment
(differential). Suppose that a free incremental charge $\delta\rho_v$ is brought into the dielectric
medium. Then the incremental work required is

$$\delta W_e = \int_v (\delta\rho_v) V \, dv \qquad (2.86)$$

where V is the electrostatic potential due to the free charges and the polarized bound
charges in the dielectric medium. Gauss's law for $\delta\rho_v$ is

$$\nabla \cdot \left(\delta\overline{D}\right) = \delta\rho_v \qquad (2.87)$$

where $\delta\overline{D}$ is the differential electric flux density. Hence (2.86) is rewritten as

$$\delta W_e = \int_v \left(\nabla \cdot \delta\overline{D}\right) V \, dv. \qquad (2.88)$$

For further simplification, we invoke the vector identity

$$\nabla \cdot \left(V \, \delta\overline{D}\right) = V \left(\nabla \cdot \delta\overline{D}\right) + \delta\overline{D} \cdot (\nabla V) \qquad (2.89)$$

and rewrite (2.88) as

$$\delta W_e = \int_v \nabla \cdot (V \, \delta \overline{D}) \; dv - \int_v \delta \overline{D} \cdot \underbrace{(\nabla V)}_{-\overline{E}} \; dv \qquad (2.90)$$

$$\underbrace{\oint_s (V \, \delta \overline{D}) \cdot d\overline{s}}$$

where the divergence theorem has been used. Let us consider field behavior when $r \to \infty$ in Fig. 2.22. The fields on a spherical surface s behave as $V \to 1/r$ and $\delta \overline{D} \to 1/r^2$ when $r \to \infty$. Since $d\overline{s}$ varies as r^2,

$$\oint_s (V \, \delta \overline{D}) \cdot d\overline{s} \to 0. \qquad (2.91)$$

Expression (2.90), therefore, becomes

$$\delta W_e = \int \delta \overline{D} \cdot \overline{E} \; dv \qquad (2.92)$$

where the integration extends over the infinite space. If the dielectric is linear such as $\overline{D} = \epsilon \overline{E}$ where ϵ is constant, then

$$\delta \overline{D} \cdot \overline{E} = \epsilon \, \delta \overline{E} \cdot \overline{E} = \frac{1}{2} \epsilon \, \delta \left(\overline{E} \cdot \overline{E} \right) = \frac{1}{2} \delta \left(\overline{D} \cdot \overline{E} \right) . \qquad (2.93)$$

Substituting (2.93) into (2.92), we obtain

$$\delta W_e = \frac{1}{2} \int \delta \left(\overline{D} \cdot \overline{E} \right) \; dv$$

$$= \frac{1}{2} \delta \left(\int \overline{D} \cdot \overline{E} \; dv \right) . \qquad (2.94)$$

Therefore the total work required to assemble the free charge from zero to ρ_v is

$$W_e = \frac{1}{2} \int \overline{D} \cdot \overline{E} \; dv. \qquad (2.95)$$

electrostatic energy stored in electric field

This is another expression of electrostatic energy stored in a linear, dielectric medium in terms of electric fields. Of course, if the medium is free space, the electrostatic energy stored in free space is

$$W_e = \frac{1}{2} \int \epsilon_0 \overline{E} \cdot \overline{E} \; dv. \qquad (2.96)$$

Fig. 2.23 Parallel-plate capacitor

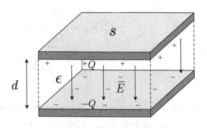

Example 2.7 Electrostatic energy stored in capacitor.

Figure 2.23 shows a parallel-plate capacitor of area s and depth d carrying charges $\pm Q$. The capacitor interior is filled with a dielectric medium of permittivity ϵ. Evaluate the energy stored in the capacitor.

Solution: If the fringing field is ignored, the electric field within the capacitor is uniform and is given by

$$E = \frac{Q}{\epsilon s}. \tag{2.97}$$

The electrostatic energy stored in the capacitor is

$$W_e = \frac{1}{2} \int \overline{D} \cdot \overline{E} \, dv$$

$$= \frac{1}{2} \epsilon \left(\frac{Q}{\epsilon s} \right)^2 sd$$

$$= \frac{1}{2} \frac{Q^2}{C} \tag{2.98}$$

where $C \left(= \dfrac{\epsilon s}{d} \right)$ is the capacitance of this capacitor. Since $Q = CV$ (V: potential difference between plates), we rewrite the energy as

$$W_e = \frac{1}{2} QV = \frac{1}{2} CV^2. \tag{2.99}$$

2.5 Poisson's Equation for Electric Fields

Laplace's and Poisson's equations are frequently used to determine electric fields in boundary-value problems. Laplace's and Poisson's equations are partial differential equations that govern potentials within a region of interest. Poisson's equation can be constructed from substituting $\overline{D} = -\epsilon \nabla V$ into Gauss's law $\nabla \cdot \overline{D} = \rho_v$ such as

$$\nabla \cdot (-\epsilon \nabla V) = \rho_v. \tag{2.100}$$

Fig. 2.24 Boundary-value problem with specified boundary condition

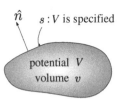

Assuming that the region is homogeneous (i.e., the medium permittivity ϵ is constant), we have

$$\nabla^2 V = -\frac{\rho_v}{\epsilon}. \qquad (2.101)$$

Poisson's equation

Furthermore, if the region contains no charge ($\rho_v = 0$), Poisson's equation is reduced to

$$\nabla^2 V = 0. \qquad (2.102)$$

Laplace's equation

Our aim is to obtain unique field solutions to these equations. Then, what does guarantee the uniqueness of field solutions? The answer lies in boundary conditions: unique solutions to these equations can be obtained by imposing appropriate boundary conditions. Figure 2.24 depicts a boundary-value problem for a potential V within a volume v that is surrounded by a surface s. Although there are different kinds of boundary conditions that make the solution unique, we will consider only the following simple, special one.

Uniqueness of the solution: The solution V to Poisson's equation is unique in v if V is specified on the boundary s.

Proof Assume that two different solutions V_1 and V_2 satisfy

$$\nabla^2 V_1 = -\frac{\rho_v}{\epsilon} \qquad (2.103)$$

$$\nabla^2 V_2 = -\frac{\rho_v}{\epsilon}. \qquad (2.104)$$

Letting $\xi = V_1 - V_2$, we have $\nabla^2 \xi = 0$. Next we utilize Green's first identity

Fig. 2.25 Electric potential
between conducting plates

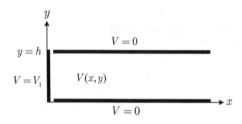

$$\oint_s \phi \nabla \psi \cdot \hat{n}\, ds = \int_v \phi \nabla^2 \psi\, dv + \int_v \nabla \phi \cdot \nabla \psi\, dv \qquad (2.105)$$

where $\hat{n}$ is a unit vector outward normal to the surface s surrounding a volume v.
Green's first identity is obtained from the divergence theorem ($\int_v \nabla \cdot \overline{A}\, dv = \oint_s \overline{A} \cdot d\overline{s}$)
with $\overline{A} = \phi \nabla \psi$. Substituting $\phi = \psi = \xi$ into (2.105), we obtain

$$\underbrace{\oint_s \xi \nabla \xi \cdot \hat{n}\, ds}_{I} = \int_v \nabla \xi \cdot \nabla \xi\, dv . \qquad (2.106)$$

Since $\xi = 0$ on s, $I = 0$ and

$$\int_v \nabla \xi \cdot \nabla \xi\, dv = 0. \qquad (2.107)$$

Since the integrand is non-negative ($\nabla \xi \cdot \nabla \xi \geq 0$), we conclude that $\nabla \xi = 0$ and ξ
should be constant over v. To determine ξ, we will again utilize the condition imposed
on s. Since $\xi = 0$ on s, $\xi = 0$ in v and $V_1 = V_2$. This proves that the solution is
unique in v if V is specified on s. In what follows, we solve some representative
boundary-value problems using Laplace's and Poisson's equations subject to the
boundary conditions.

Example 2.8 Electric potential between conducting parallel plates.
Figure 2.25 shows two semi-infinite parallel conducting plates at zero potential separated by a distance h. A third conducting plate at a potential V_1 is placed at $x = 0$.
Assume that the electric potential is a function of x and y. Determine the electric
field in the region surrounded by these three conducting plates.

Solution: The potential V within the region ($0 < y < h$ and $x > 0$) satisfies
Laplace's equation:

$$\frac{\partial^2 V}{\partial x^2} + \frac{\partial^2 V}{\partial y^2} = 0. \qquad (2.108)$$

We assume the solution in the product form of

$$V = XY \tag{2.109}$$

where X and Y are functions of x and y, respectively. Substituting (2.109) into (2.108) and dividing (2.108) by XY, we obtain

$$\underbrace{\frac{1}{X}\frac{d^2 X}{dx^2}}_{\text{function of } x} + \underbrace{\frac{1}{Y}\frac{d^2 Y}{dy^2}}_{\text{function of } y} = 0. \tag{2.110}$$

The first and second terms must be equal to the constants $\pm p^2$ irrespective of x and y as

$$\frac{1}{X}\frac{d^2 X}{dx^2} = p^2 \tag{2.111}$$

$$\frac{1}{Y}\frac{d^2 Y}{dy^2} = - p^2. \tag{2.112}$$

The solutions are

$$\frac{d^2 X}{dx^2} - p^2 X = 0 \Longrightarrow X = c_1 e^{px} + c_2 e^{-px} \tag{2.113}$$

$$\frac{d^2 Y}{dy^2} + p^2 Y = 0 \Longrightarrow Y = c_3 \sin py + c_4 \cos py \tag{2.114}$$

where c_1 through c_4 are unknown coefficients to be determined.

1. The boundary condition $V|_{y=0} = 0$ yields $c_4 = 0$. Hence

$$Y = c_3 \sin py. \tag{2.115}$$

2. The boundary condition $V|_{y=h} = 0$ yields $p = n\pi/h$ for $n = 1, 2, 3 \cdots$.
3. The boundary condition $V|_{x=\infty} = 0$ yields $c_1 = 0$ and

$$V = c_2 c_3 \exp\left(-\frac{n\pi}{h}x\right) \sin\left(\frac{n\pi}{h}y\right). \tag{2.116}$$

A complete solution is written as a superposition of (2.116) as

$$V = \sum_{n=1}^{\infty} C_n \exp\left(-\frac{n\pi}{h}x\right) \sin\left(\frac{n\pi}{h}y\right). \tag{2.117}$$

4. The boundary condition of $V|_{x=0} = V_1$ for $0 \le y \le h$ yields

$$V_1 = \sum_{n=1}^{\infty} C_n \sin\left(\frac{n\pi}{h}y\right) \tag{2.118}$$

which is a Fourier sine series representation of V_1 over the interval $0 \le y \le h$. To determine the coefficient C_n, we use the orthogonal property of sinusoidal functions as

$$\int_0^h \sin\left(\frac{n\pi}{h}y\right)\sin\left(\frac{m\pi}{h}y\right)dy = \begin{cases} \dfrac{h}{2} & \text{for } m = n \\ 0 & \text{for } m \ne n \end{cases} \tag{2.119}$$

where m and n are $1, 2, 3, \cdots$. Multiplying (2.118) by $\sin\left(\dfrac{m\pi}{h}y\right)$ and integrating it with respect to y from 0 to h, we obtain

$$V_1 \underbrace{\int_0^h \sin\left(\frac{m\pi}{h}y\right)dy}_{\dfrac{h[1-(-1)^m]}{m\pi}} = \underbrace{\sum_{n=1}^{\infty} C_n \int_0^h \sin\left(\frac{n\pi}{h}y\right)\sin\left(\frac{m\pi}{h}y\right)dy}_{\dfrac{h}{2}C_m}. \tag{2.120}$$

Therefore

$$V = \sum_{n=1,3,5,\cdots}^{\infty} \frac{4V_1}{n\pi} \exp\left(-\frac{n\pi}{h}x\right)\sin\left(\frac{n\pi}{h}y\right). \tag{2.121}$$

The electric field is

$$\begin{aligned} \overline{E} &= -\nabla V \\ &= \frac{4V_1}{h}\sum_{n=1,3,5,\cdots}^{\infty} \exp\left(-\frac{n\pi}{h}x\right)\left[\hat{x}\sin\left(\frac{n\pi}{h}y\right) - \hat{y}\cos\left(\frac{n\pi}{h}y\right)\right]. \end{aligned} \tag{2.122}$$

Example 2.9 Electric potential due to a charged sphere.
Figure 2.26 shows a charged sphere of radius a with a uniform volume charge density ρ_v. Evaluate the electric field produced by the sphere.

Solution: Due to spherical symmetry, the electric field has the radial component E_r and the electric potential V is dependent on r only. Poisson's equation is written as

$$\frac{1}{r^2}\frac{d}{dr}\left(r^2\frac{dV}{dr}\right) = \begin{cases} 0 & \text{for } r > a \\ -\dfrac{\rho_v}{\epsilon_0} & \text{for } r \le a. \end{cases} \tag{2.123}$$

Fig. 2.26 A spherical charge with a uniform charge distribution ρ_v

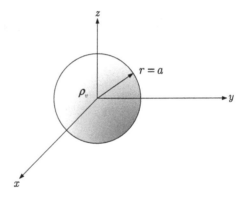

For the region $r > a$

$$\frac{1}{r^2}\frac{d}{dr}\left(r^2\frac{dV_1}{dr}\right) = 0 \Longrightarrow \frac{dV_1}{dr} = \frac{c_1}{r^2}. \tag{2.124}$$

where c_1 is the unknown coefficient. For the charged region $r \le a$

$$\frac{1}{r^2}\frac{d}{dr}\left(r^2\frac{dV_2}{dr}\right) = -\frac{\rho_v}{\epsilon_0} \Longrightarrow \frac{dV_2}{dr} = -\frac{\rho_v}{3\epsilon_0}r + \frac{c_2}{r^2}. \tag{2.125}$$

where c_2 is the unknown coefficient. The following two boundary conditions are needed to determine c_1 and c_2:

1. The electric field at $r = 0$ is zero due to spherical symmetry.

$$E_r\big|_{r=0} = -\frac{dV_2}{dr}\bigg|_{r=0} = 0 \Longrightarrow c_2 = 0. \tag{2.126}$$

2. The electric field is continuous at $r = a$, yielding

$$-\frac{dV_1}{dr}\bigg|_{r=a} = -\frac{dV_2}{dr}\bigg|_{r=a} \Longrightarrow c_1 = -\frac{\rho_v}{3\epsilon_0}a^3. \tag{2.127}$$

The electric field is

$$E_r = \begin{cases} \dfrac{\rho_v a^3}{3\epsilon_0 r^2} & \text{for } r > a \\[2mm] \dfrac{\rho_v}{3\epsilon_0}r & \text{for } r \le a. \end{cases} \tag{2.128}$$

Note that (2.128) is identical with the solution derived from Gauss's law in Example 2.3.

Fig. 2.27 A charge q above
a conducting plane of infinite
extent

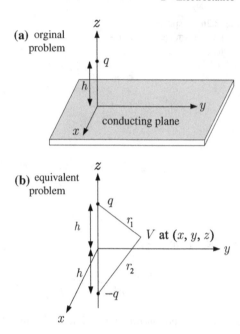

2.6 Image Method

The image method is often used to determine fields when charges are placed nearby
conducting planes. The image method enables us to solve boundary-value problems
associated with conducting planes in a simple manner. For illustration, we consider a
point charge q placed above a conducting plane, as shown in Fig. 2.27a. The conducting plane is infinitely extended at $z = 0$ and its electric scalar potential is assumed
zero. To determine the electric field for $z > 0$, we replace the original problem
(a) by the equivalent problem (b), where the conducting plane was removed and an
image charge $-q$ was placed at $z = -h$. The inclusion of $-q$ at $z = -h$ makes the
electric potential at $z = 0$ zero, thereby making the two problems equivalent as far
as the region $z \geq 0$ is concerned. In other words, the boundary condition at $z = 0$
is satisfied and the electric potential for $z > 0$ in problem (b) is identical with that
in problem (a). The evaluation of electric potential in problem (b) is straightforward
since the conducting plane was removed. The electric potential at a position (x, y, z)
due to q and $-q$ is given by the sum of two point charge responses as

$$V = \frac{q}{4\pi\epsilon_0} \left(\frac{1}{r_1} - \frac{1}{r_2} \right)$$

$$= \frac{q}{4\pi\epsilon_0} \left\{ \frac{1}{[x^2 + y^2 + (z-h)^2]^{1/2}} \right.$$

$$-\frac{1}{[x^2 + y^2 + (z+h)^2]^{1/2}}\Bigg\}. \tag{2.129}$$

Expression (2.129) represents a formal solution to problem (a). The electric field at (x, y, z) is

$$\overline{E} = -\nabla V$$

$$= \frac{q}{4\pi\epsilon_0} \Bigg\{ \frac{\hat{x}x + \hat{y}y + \hat{z}(z-h)}{[x^2 + y^2 + (z-h)^2]^{3/2}}$$

$$-\frac{\hat{x}x + \hat{y}y + \hat{z}(z+h)}{[x^2 + y^2 + (z+h)^2]^{3/2}} \Bigg\}. \tag{2.130}$$

The electric field on the conducting surface at $z = 0$ is shown to be

$$\overline{E}\Big|_{z=0} = \hat{z} \underbrace{\frac{(-qh)}{2\pi\epsilon_0 \left(x^2 + y^2 + h^2\right)^{3/2}}}_{E_n} \tag{2.131}$$

which is normal to the conducting plane. The surface charge density ρ_s $(= \epsilon_0 E_n)$ is induced on the conducting plane. The total charge q_i induced on the conducting plane is

$$q_i = \int_{-\infty}^{\infty} \int_{-\infty}^{\infty} \rho_s \, dx \, dy$$

$$= \int_{-\infty}^{\infty} \int_{-\infty}^{\infty} \frac{-qh}{2\pi(x^2 + y^2 + h^2)^{3/2}} \, dx \, dy. \tag{2.132}$$

To perform integration, we introduce the polar coordinates (ρ, ϕ) such as

$$x = \rho \cos \phi \tag{2.133}$$

$$y = \rho \sin \phi. \tag{2.134}$$

Then

$$q_i = \int_0^{2\pi} \int_0^{\infty} \frac{-qh}{2\pi(\rho^2 + h^2)^{3/2}} \rho \, d\rho \, d\phi$$

$$= \int_0^{\infty} \frac{-qh}{(\rho^2 + h^2)^{3/2}} \rho \, d\rho$$

$$= -q. \tag{2.135}$$

This implies that the electric flux emanating from the real charge q terminates at the conducting plane as if the image charge $-q$ were present at $z = -h$.

2.7 Problems for Chapter 2

1. Figure 2.28 shows an infinitesimally thin circular disk charge of surface charge density ρ_s in free space. Evaluate the electric field on the z-axis.
   Hint: The electric field on the z-axis has a component in the z-direction.
2. Figure 2.29 depicts a long cylindrical charge with a radius a placed in free space of permittivity ϵ_0. A charge Q per unit length is uniformly distributed within a circular cylinder. Determine the electric field.
   Hint: Construct a cylindrical Gaussian surface surrounding the charge and apply Gauss's law.
3. Two slab charges of thickness a with volume charge densities $\pm\rho_v$ are shown in Fig. 2.30. Assume that the slabs are infinitely extended in the y- and z-directions. Evaluate the electric field versus x.
   Hint: There is no field variation in the y- and z-directions. The electric field has an x-component. Use Gauss's law.
4. Figure 2.31 shows a plane boundary between medium 1 (permittivity: ϵ_1) and medium 2 (permittivity: ϵ_2). There is no free surface charge on the boundary. Determine $\overline{E}_2\big|_{y=0}$ in medium 2 if the electric field in medium 1 is $\overline{E}_1\big|_{y=0} = \hat{x}E_x + \hat{y}E_y$.
   Hint: Use the boundary conditions $D_{1n} = D_{2n}$ and $E_{1t} = E_{2t}$.
5. Figure 2.32 shows two circular conducting cylinders of infinite length and radius a separated by a spacing d $(d \gg a)$ in air. Determine the capacitance per unit length.
   Hint: Assume a uniform charge distribution and apply Gauss's law.
6. Consider a charge sphere of radius a with a volume charge density ρ_v in free space. Calculate the electrostatic energy required to assemble this spherical charge.
7. Figure 2.33 illustrates a conducting solid sphere of radius a carrying a charge Q. Find the electrostatic potential inside and outside the sphere.
   Hint: Charges exist on the surface of a sphere. Use Gauss's law to determine the electric field and the electrostatic potential.

Fig. 2.28 Circular disk charge of radius a

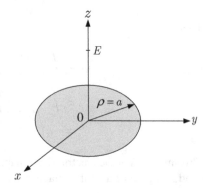

Fig. 2.29 A long cylindrical charge

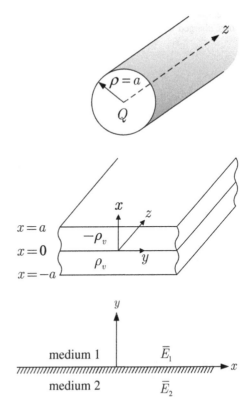

Fig. 2.30 Two slab charges of infinite extent

Fig. 2.31 Two dielectric media

Fig. 2.32 Charged two conducting cylinders in parallel

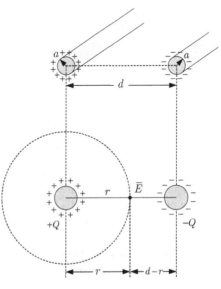

Fig. 2.33 A conducting
sphere charged with Q

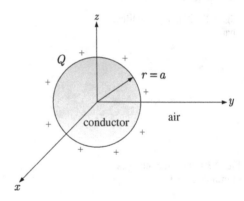

Fig. 2.34 A coaxial capacitor
of length $l \gg b$

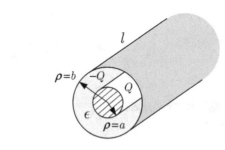

Fig. 2.35 Electric potential
within conducting plates

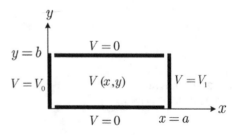

8. Figure 2.34 shows a coaxial capacitor of length l filled with a dielectric of ϵ.
   Evaluate the energy stored between the inner and outer conductors carrying $+Q$
   and $-Q$ charges, respectively.
   Hint: Use Gauss's law to find the electric field.

9. Consider four conducting plates in Fig. 2.35. Two parallel conducting plates of
   zero potential are placed at $y = 0$ and b. A third conducting plate at a potential
   V_0 is placed at $x = 0$ and a fourth conducting plate at a potential V_1 is placed
   at $x = a$. Assume that the electric scalar potential is a function of x and y.
   Determine the electric scalar potential in the rectangular region surrounded by
   these four conducting plates.
   Hint: Use Laplace's equation.

Fig. 2.36 An infinitely long
line charge

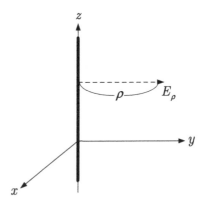

10. Figure 2.36 shows an infinitely long line charge with a uniform line charge density ρ_l in air. Evaluate the electric field by using the following three different approaches:

- electric field expression

$$\overline{E} = \int_{v'} \hat{R} \frac{\rho'_v}{4\pi\epsilon_0 R^2} \, dv' \tag{2.136}$$

- potential expression

$$V = \int_{v'} \frac{\rho'_v}{4\pi\epsilon_0 R} \, dv' \tag{2.137}$$

- Gauss's law

$$\oint_s \overline{E} \cdot d\overline{s} = \frac{Q}{\epsilon_0}. \tag{2.138}$$

Chapter 3
Magnetostatics

3.1 Conduction Currents

3.1.1 Steady Currents

Moving charges are usually called currents. Specifically, when a charge with a volume charge density ρ_v moves at a velocity $\bar{u}$, this moving charge is characterized by the current density (or volume current density) of (Fig. 3.1)

$$\bar{J} = \rho_v \bar{u}. \tag{3.1}$$

Consider the current density $\bar{J}$ in Fig. 3.2. The total current I passing through an area s is given by

$$I = \int_s \bar{J} \cdot d\bar{s} \tag{3.2}$$

where the direction of $\bar{s}$ is normal to the surface s. Figure 3.3 illustrates a steady current that flows through a conductor under the influence of electric field. When a voltage (equivalently static electric field $\bar{E}$) is applied across a conductor, free electrons in a conductor start to move, attain speed, and maintain a current I. The steady current I enters the conductor, exits the conductor, and leaves no static charges behind. In order to quantify the steady current in terms of a current density, we will consider a small volume v in Fig. 3.3. A current entering v equals a current exiting v. In other words, the total net current I_{net} flowing through a surface s enclosing a volume v must be zero:

$$I_{net} = \oint_s \bar{J} \cdot d\bar{s} = 0. \tag{3.3}$$

H. J. Eom, *Primary Theory of Electromagnetics*, Power Systems,
DOI: 10.1007/978-94-007-7143-7_3, © Springer Science+Business Media Dordrecht 2013

Magnetostatics

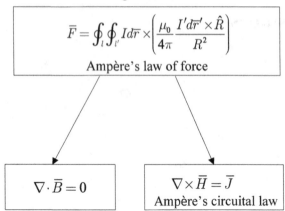

$$\bar{F} = \oint_l \oint_{l'} I d\bar{r} \times \left(\frac{\mu_0}{4\pi} \frac{I' d\bar{r}' \times \hat{R}}{R^2} \right)$$

Ampère's law of force

$$\nabla \cdot \bar{B} = 0$$

$$\nabla \times \bar{H} = \bar{J}$$
Ampère's circuital law

Fig. 3.1 Fundamental equations for static magnetic fields

Fig. 3.2 Current density $\bar{J}$
passing through s

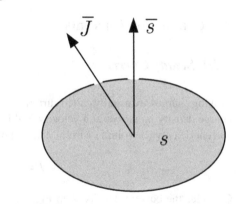

Fig. 3.3 Steady current in a
conductor

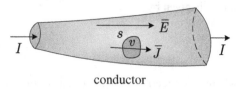

conductor

Here $\oint_s$ denotes integration over the closed surface s where the vector $d\bar{s}$ points away from v. Expression (3.3) is referred to as Kirchhoff's current law in circuit theory: the summation of currents flowing out of a node must be zero. Applying the divergence theorem to (3.3) yields

$$\int_v \nabla \cdot \bar{J} \, dv = 0. \tag{3.4}$$

Since (3.4) is valid irrespective of v, the steady current is characterized by

$$\nabla \cdot \overline{J} = 0. \tag{3.5}$$

steady current density

3.1.2 Ohm's Law

We again consider a conductor under the influence of electric field in Fig. 3.3. When a voltage (or equivalently electric field $\overline{E}$) is applied across the conductor, free electrons experience a force and start to move. Moving electrons are bound to collide with atomic lattices within the conductor, thereby attaining a constant terminal velocity $\overline{u}$ that is called a drift velocity. The terminal velocity $\overline{u}$ is proportional to the applied electric field $\overline{E}$ and is expressed by

$$\overline{u} = -\mu_e \overline{E} \tag{3.6}$$

where the proportionality constant μ_e is the electron mobility. The electron mobility for copper is about 3.2×10^{-3} $(\mathrm{m^2\,V^{-1}\,s^{-1}})$, indicating that the electron drift velocity in good conductors is rather slow. Substituting (3.6) into (3.1) gives

$$\overline{J} = \sigma \overline{E} \tag{3.7}$$

Ohm's law

where $\sigma = -\rho_v \mu_e$ is the conductivity of a conductor. The reciprocal of the conductivity is called the resistivity. The current flowing through a conductor wire is called the conduction current. Ohm's law was established on experimental observations of conduction currents. Examples of conductors are copper, aluminium, gold, etc.; the conductivity of copper is $\sigma = 5.8 \times 10^7$ (S/m). An idealized conductor whose conductivity is infinite ($\sigma \to \infty$) is referred to as a perfect conductor (or perfect electric conductor, PEC for short). The perfect conductor, although an idealized material, is a very useful concept in many practical situations.

Example 3.1 Ohm's law for a conducting wire.
Figure 3.4 shows a cylindrical conductor (metallic wire) of length l, cross-sectional area s, and conductivity σ. Derive Ohm's law expression used in circuit theory.

Solution: Assume that an electric field $\overline{E}$ is applied across a cylindrical conductor. According to Ohm's law a conduction current I is

Fig. 3.4 Cylindrical conductor under the influence of
electric field $\overline{E}$

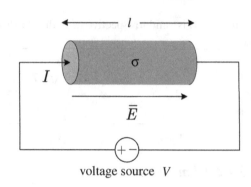

$$\overline{J} = \sigma\overline{E} \implies \underbrace{\int_{s} \overline{J} \cdot d\overline{s}}_{I} = \underbrace{\int_{s} \sigma\overline{E} \cdot d\overline{s}}_{\sigma E s}. \tag{3.8}$$

The potential difference across the conductor is

$$V = -\int \overline{E} \cdot d\overline{r} = El. \tag{3.9}$$

Expressions (3.8) and (3.9) give Ohm's law in circuit theory as

$$V = \underbrace{\frac{l}{\sigma s}}_{R} I \tag{3.10}$$

where R is called the resistance of a conductor. Ohm's law states that the current
flowing in a metallic wire is proportional to the potential difference applied across
the ends of wire.

Example 3.2 A spherical capacitor having an imperfect dielectric.

Figure 3.5 shows a spherical capacitor consisting of two concentric conducting
spherical surfaces of radii a and b. The space between the two surfaces is filled
with a homogeneous, lossy dielectric with a conductivity σ. When a voltage source
$V_0 \, (= V|_{r=a} - V|_{r=b})$ is applied across the two conducting surfaces, determine
the leakage current I and the leakage resistance R between the two surfaces.

Solution: The conduction current density $\overline{J}$ in a conductive medium obeys
Ohm's law

$$\overline{J} = \sigma\overline{E} = -\sigma\nabla V \tag{3.11}$$

and the steady current condition

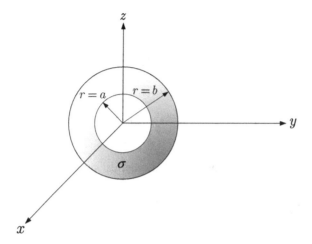

Fig. 3.5 A spherical capacitor having concentric spherical surfaces of radii a and b

$$\nabla \cdot \overline{J} = 0. \tag{3.12}$$

Substitution of (3.11) into (3.12) gives Laplace's equation

$$\nabla^2 V = 0. \tag{3.13}$$

Here the conductivity σ has been assumed to be constant. Due to spherical symmetry, the electric field has a radial r-component and the electric potential V is dependent on r only. Hence, Laplace's equation for the region $a \le r \le b$ is rewritten as

$$\frac{1}{r^2}\frac{d}{dr}\left(r^2\frac{dV}{dr}\right) = 0. \tag{3.14}$$

The electric scalar potential is written as

$$V = -\frac{c_1}{r} + c_2. \tag{3.15}$$

Since $V_0 = V\big|_{r=a} - V\big|_{r=b}$, the unknown coefficient c_1 is

$$c_1 = -\frac{abV_0}{b-a}. \tag{3.16}$$

Thus the potential is rewritten as

$$V = \frac{ab}{b-a}\frac{V_0}{r} + c_2. \tag{3.17}$$

From V we calculate I and R subsequently as follows:

$$\overline{E} = -\nabla V = \hat{r}\frac{ab}{b-a}\frac{V_0}{r^2}. \tag{3.18}$$

From Ohm's law

$$\overline{J} = \sigma\overline{E} = \hat{r}\frac{\sigma ab}{b-a}\frac{V_0}{r^2}. \tag{3.19}$$

The total leakage current flowing through a spherical surface with a radius r is

$$I = \int_s \overline{J} \cdot d\overline{s}$$

$$= \frac{4\pi\sigma ab}{b-a}V_0. \tag{3.20}$$

Hence the leakage resistance is

$$R = \frac{V_0}{I} = \frac{b-a}{4\pi\sigma ab}. \tag{3.21}$$

3.1.3 Joule Heating

When an electric field $\overline{E}$ is applied across a conductor, a differential volume dv with a charge density ρ_v experiences a differential force $d\overline{F}$ $(= \rho_v \, dv \, \overline{E})$ and attains a velocity $\overline{u}$. We note that an infinitesimal work d^2W is required to displace a charge $\rho_v \, dv$ by a distance $d\overline{r}$ as

$$d^2W = d\overline{F} \cdot d\overline{r} = \rho_v \, dv \, \overline{E} \cdot d\overline{r}. \tag{3.22}$$

The differential work to displace charges that occupy a conductor volume v is

$$dW = \int_v \left(\rho_v\overline{E} \cdot d\overline{r}\right) dv. \tag{3.23}$$

The work done by moving charges is dissipated into heat due to collisions with atoms. Power is defined as the rate of work per unit time. The total power dissipated in Joule heat by a conductor occupying a volume v is

$$P = \frac{dW}{dt} = \int_v \rho_v \left(\overline{E} \cdot \frac{d\overline{r}}{dt}\right) dv. \tag{3.24}$$

Note

$$\rho_v \frac{d\bar{r}}{dt} = \rho_v \bar{u} = \bar{J}.$$ (3.25)

Due to Ohm's law $\bar{E} = \bar{J}/\sigma$, (3.24) is rewritten as

$$P = \int_v \frac{J^2}{\sigma} dv$$ (3.26)

where $J = |\bar{J}|$. The above Joule-heating formula enables us to calculate the power dissipation by current-carrying conductors.

Example 3.3 Ohmic power loss.
Consider a cylindrical conductor in Fig. 3.4. When a voltage V is applied across the conductor of resistance R, a current I flows through the conductor. Express the power loss in terms of I and R.

Solution: The Joule heat loss is

$$P = \int_v \frac{J^2}{\sigma} dv = \frac{J^2}{\sigma} sl.$$ (3.27)

Since $J = \frac{I}{s}$ and $R = \frac{l}{\sigma s}$, we obtain

$$P = I^2 R$$ (3.28)

which is a familiar expression of ohmic power loss used in circuit theory.

3.2 Fundamentals of Magnetic Fields

Currents are known to exert magnetic forces on each other. Magnetic forces are discussed in terms of magnetic flux densities or equivalently magnetic fields. Formulas for evaluating magnetic flux densities from currents include Biot-Savart law, Ampère's circuital law, and magnetic vector potential. We will show that all these formulas can be derived from Ampère's law of force.

3.2.1 Ampère's Law of Force and Magnetic Flux Density

In Fig. 3.6, we consider two time-invariant current loops in free space. The locations of current loops are designated by the position vectors $\bar{r}$ and $\bar{r}'$. The currents I and

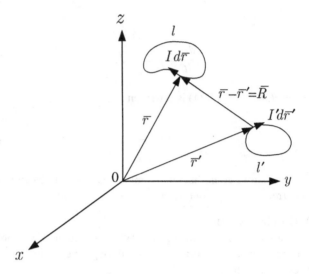

Fig. 3.6 Two current loops at $\bar{r}$ and $\bar{r}'$

I' exist in the form of current loops l and l', respectively. In other words, I and I' are steady currents flowing around the loops. Two current-carrying loops l and l' exert magnetic force on each other. A differential current segment $I'd\bar{r}'$ exerts an infinitesimal force $d^2\bar{F}$ on a differential current segment $Id\bar{r}$ as

$$d^2\bar{F} = \frac{\mu_0}{4\pi} \frac{Id\bar{r} \times \left(I'd\bar{r}' \times \hat{R}\right)}{R^2} \tag{3.29}$$

where $\hat{R}$ is the unit vector ($\hat{R} = \bar{R}/R$) and μ_0 is the permeability of free space (vacuum) as

$$\mu_0 = 4\pi \times 10^{-7} \text{ (H/m)}. \tag{3.30}$$

The meaning of $d^2\bar{F}$ is such as

$$d\bar{F} = \oint_{l'} d^2\bar{F} \tag{3.31}$$

where the symbol $\oint_{l'}$ denotes the line integral over the closed loop l'. Here $d\bar{F}$ refers to a differential force that I' exerts on a differential current segment $Id\bar{r}$ as

$$d\overline{F} = I d\overline{r} \times \underbrace{\left(\frac{\mu_0}{4\pi} \oint_{l'} \frac{I' d\overline{r}' \times \hat{R}}{R^2} \right)}_{\overline{B}}. \tag{3.32}$$

The notation $\overline{B}$ is called a magnetic flux density, which is produced by the current I'. This relation between $\overline{B}$ and I' is known as the Biot-Savart law:

$$\overline{B} = \frac{\mu_0}{4\pi} \oint_{l'} \frac{I' d\overline{r}' \times \hat{R}}{R^2}. \tag{3.33}$$

Biot-Savart law

The Biot-Savart law enables us to evaluate the magnetic flux density at $\overline{r}$ due to the loop current at $\overline{r}'$. We further evaluate the force $\overline{F} \; (= \oint_l d\overline{F})$ exerted on the loop l due to the loop l' as

$$\overline{F} = \frac{\mu_0 I I'}{4\pi} \oint_l \oint_{l'} \frac{d\overline{r} \times (d\overline{r}' \times \hat{R})}{R^2}. \tag{3.34}$$

Ampère's law of force

Ampère's law of force and the Biot-Savart law are closely related to each other. It is possible to rewrite Ampère's law of force in slightly different form. Since

$$d\overline{r} \times (d\overline{r}' \times \hat{R}) = d\overline{r}'(d\overline{r} \cdot \hat{R}) - \hat{R}(d\overline{r} \cdot d\overline{r}') \tag{3.35}$$

we have

$$\overline{F} = \frac{\mu_0 I I'}{4\pi} \left[\oint_{l'} d\overline{r}' \oint_l \frac{d\overline{r} \cdot \hat{R}}{R^2} - \oint_l \oint_{l'} \frac{\hat{R}(d\overline{r} \cdot d\overline{r}')}{R^2} \right]. \tag{3.36}$$

We note

$$\oint_l \frac{d\overline{r} \cdot \hat{R}}{R^2} = \int_s \underbrace{\left(\nabla \times \frac{\hat{R}}{R^2} \right)}_{0} \cdot d\overline{s} = 0 \tag{3.37}$$

where Stokes's theorem has been used. Hence we obtain

Fig. 3.7 A steady volume
current density loop $\overline{J}'$

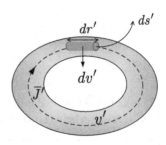

$$\overline{F} = -\frac{\mu_0 I I'}{4\pi} \oint_l \oint_{l'} \frac{\hat{R}(d\overline{r} \cdot d\overline{r}')}{R^2} \tag{3.38}$$

which is known as Neumann's formula for magnetic forces. Note that interchanging l and l' in (3.38) produces a negative sign stemming from $\hat{R} \left(= \dfrac{\overline{r} - \overline{r}'}{|\overline{r} - \overline{r}'|} \right)$. Therefore Ampère's law of force (3.38) satisfies

$$\overline{F} = -\overline{F'} \tag{3.39}$$

where $\overline{F'}$ denotes the force exerted on the loop l' due to the loop l carrying a current I. Equation (3.39) obeys Newton's third law, which states that action and reaction are equal but opposite.

Biot-Savart law for volume currents: Although the Biot-Savart law was given in the form of line currents, it is possible to extend the Biot-Savart law to other types of currents such as volume currents. Figure 3.7 illustrates a steady volume current density loop $\overline{J}'$ that occupies a volume v'. Consider a differential volume dv' ($= dr' \, ds'$). Then the differential current density element $\overline{J}' \, dv'$ is

$$\overline{J}' \, dv' = \overline{J}' \, ds' \, dr' = I' \, d\overline{r}' \tag{3.40}$$

where I' is the current flowing through ds'. Note that the current element $(\overline{J}' \, dv')$ in volume currents is equivalent to the current element $(I' \, d\overline{r}')$ in line currents. Therefore it is possible to replace $I' \, d\overline{r}'$ with $\overline{J}' \, dv'$ in the Biot-Savart law. The result is

$$\overline{B} = \frac{\mu_0}{4\pi} \int_{v'} \frac{\overline{J}' \times \hat{R}}{R^2} \, dv'. \tag{3.41}$$

Equation (3.41) is referred to as the Biot-Savart law for voluminous currents occupying a volume v'.

Fig. 3.8 Magnetic vector potential due to a current loop

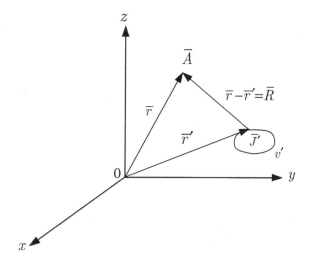

3.2.2 Magnetic Vector Potential

Besides the Biot-Savart law, there is an alternative, efficient approach that allows us to evaluate the magnetic flux density. This alternative approach is based on a magnetic vector potential and can be introduced from (3.41). Consider a current loop $\overline{J}'$ in Fig. 3.8 and we shall become a little foresighted as

$$\nabla \times \left(\frac{\overline{J}'}{R}\right) = \frac{1}{R}\underbrace{\left(\nabla \times \overline{J}'\right)}_{0} - \overline{J}' \times \underbrace{\left(\nabla \frac{1}{R}\right)}_{-\hat{R}/R^2}$$

$$= \frac{\overline{J}' \times \hat{R}}{R^2}. \tag{3.42}$$

Consider (3.41) in view of (3.42). Since the symbol $\int_{v'}$ is with respect to (x', y', z') and ∇ is with respect to (x, y, z), the order of integration and differentiation in (3.41) is interchangeable. Hence

$$\overline{B} = \nabla \times \frac{\mu_0}{4\pi} \int_{v'} \frac{\overline{J}'}{R} \, dv'. \tag{3.43}$$

From (3.43) we introduce the magnetic vector potential (or simply vector potential) $\overline{A}$ as

$$\overline{A} = \frac{\mu_0}{4\pi} \int_{v'} \frac{\overline{J'}}{R} \, dv'. \qquad (3.44)$$

definition of magnetic vector potential

Specifically $\overline{J'}$ is a function of (x', y', z') and $\overline{A}$ is a function of (x, y, z), where

$$R = \sqrt{(x - x')^2 + (y - y')^2 + (z - z')^2} \qquad (3.45)$$

$$dv' = dx' \, dy' \, dz'. \qquad (3.46)$$

The magnetic vector potential at (x, y, z) is obtained by summing up the contributions of current density $\overline{J'}$ over (x', y', z'). Then the magnetic flux density is rewritten as

$$\overline{B} = \nabla \times \overline{A}. \qquad (3.47)$$

magnetic flux density in terms of $\overline{A}$

The vector identity $\nabla \cdot \left(\nabla \times \overline{A} \right) \equiv 0$ leads to

$$\nabla \cdot \overline{B} = 0. \qquad (3.48)$$

magnetic Gauss's law

Applying the divergence theorem to the above expression yields

$$\oint_{s} \overline{B} \cdot d\overline{s} = 0 \qquad (3.49)$$

magnetic Gauss's law in integral form

which shows that the magnetic flux through any closed surface s is zero.

Poisson's equation: In order to evaluate $\overline{B}$ from $\overline{A}$, we need an additional equation of $\overline{A}$. Poisson's equation is a differential equation governing $\overline{A}$. To derive Poisson's equation, we consider the Laplacian of A_x as

$$\nabla^2 A_x = \frac{\mu_0}{4\pi} \nabla^2 \left(\int_{v'} \frac{J_x'}{R} \, dv' \right) \tag{3.50}$$

where J_x' is the x-component of $\overline{J}'$. Since J_x' is a function of (x', y', z') and ∇^2 operates with respect to (x, y, z), we have

$$\nabla^2 A_x = \frac{\mu_0}{4\pi} \int_{v'} J_x' \nabla^2 \left(\frac{1}{R} \right) dv'. \tag{3.51}$$

Furthermore we note

$$\nabla^2 \left(\frac{1}{R} \right) = \nabla \cdot \nabla \left(\frac{1}{R} \right) = -\nabla \cdot \left(\frac{\hat{R}}{R^2} \right) = -4\pi\delta(\overline{R}). \tag{3.52}$$

The relation $\nabla \cdot \left(\frac{\hat{R}}{R^2} \right) = 4\pi\delta(\overline{R})$ was derived in Example 1.1, where $\overline{R} = \overline{r} - \overline{r}'$ and $\delta(\overline{R})$ is the Dirac delta function. We rewrite (3.51) as

$$\nabla^2 A_x = -\mu_0 \int_{v'} J_x' \, \delta(\overline{R}) \, dv'$$

$$= -\mu_0 J_x \tag{3.53}$$

where J_x is a function of $\overline{r}$ as $J_x = J_x' \big|_{\overline{r}' \to \overline{r}}$. Similarly we obtain

$$\nabla^2 A_y = -\mu_0 J_y \tag{3.54}$$
$$\nabla^2 A_z = -\mu_0 J_z. \tag{3.55}$$

The combination of (3.53–3.55) yields a vector form

$$\nabla^2 \overline{A} = -\mu_0 \overline{J} \tag{3.56}$$

Poisson's equation

where $\overline{J} = \overline{J}' \big|_{\overline{r}' \to \overline{r}}$. Note that both $\overline{J}$ and $\overline{A}$ are functions of $\overline{r}$. Poisson's equation subject to prescribed boundary conditions is used to determine the magnetic fields in boundary-value problems.

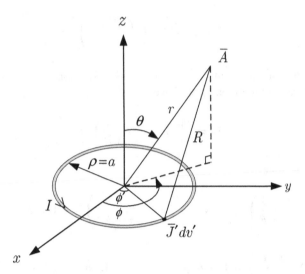

Fig. 3.9 Magnetic field due to a circular current loop of radius a (magnetic dipole)

Example 3.4 Magnetic field due to a magnetic dipole.
A circular current loop with a radius a at the origin carries a current I, as shown in Fig. 3.9. A small current loop is called a magnetic dipole. Evaluate the magnetic flux density from the magnetic dipole when $r \gg a$.

Solution: It is expedient to work with the rectangular coordinates (x, y, z) for analysis. The primed and unprimed coordinates refer to the source $(\overline{J}')$ location and the observation $(\overline{A})$ location, respectively. The differential current is

$$\overline{J}' \, dv' = I a d\phi' \underbrace{\left(-\hat{x} \sin \phi' + \hat{y} \cos \phi'\right)}_{\hat{\phi}'}. \tag{3.57}$$

The magnetic vector potential is given by

$$A_x = -\frac{\mu_0 I a}{4\pi} \int_0^{2\pi} \frac{\sin \phi'}{R} \, d\phi' \tag{3.58}$$

$$A_y = \frac{\mu_0 I a}{4\pi} \int_0^{2\pi} \frac{\cos \phi'}{R} \, d\phi'. \tag{3.59}$$

To determine R, we consider a plane triangle in Fig. 3.10 where

$$\overline{r} = r(\hat{x} \sin \theta \cos \phi + \hat{y} \sin \theta \sin \phi + \hat{z} \cos \theta) \tag{3.60}$$

$$\overline{r}' = a(\hat{x} \cos \phi' + \hat{y} \sin \phi') \tag{3.61}$$

$$\overline{r} \cdot \overline{r}' = ra \underbrace{\sin \theta \cos(\phi - \phi')}_{\cos \psi}. \tag{3.62}$$

Fig. 3.10 A triangle with sides r, R, and r'

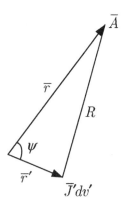

Furthermore, law of cosines gives

$$
\begin{aligned}
R^2 &= r^2 + a^2 - 2ar\cos\psi \\
&= r^2 + a^2 - 2ar\sin\theta\cos(\phi - \phi').
\end{aligned}
\tag{3.63}
$$

For $r \gg a$,

$$
\begin{aligned}
\frac{1}{R} &= \left[r^2 + a^2 - 2ar\sin\theta\cos(\phi - \phi') \right]^{-0.5} \\
&\approx \left[r^2 - 2ar\sin\theta\cos(\phi - \phi') \right]^{-0.5} \\
&\approx \frac{1}{r}\left[1 - \underbrace{\frac{2a}{r}\sin\theta\cos(\phi - \phi')}_{\xi} \right]^{-0.5}.
\end{aligned}
\tag{3.64}
$$

Note that the binomial series for $|\xi| < 1$ is given by

$$
\begin{aligned}
(1 - \xi)^n &= 1 - n\xi + \frac{n(n-1)}{2!}\xi^2 - \cdots \\
&\approx 1 - n\xi.
\end{aligned}
\tag{3.65}
$$

Therefore

$$
\frac{1}{R} \approx \frac{1}{r} + \frac{a}{r^2}\sin\theta\cos(\phi - \phi').
\tag{3.66}
$$

Substituting (3.66) into (3.58) and (3.59), and performing integration, we obtain

$$A_x \approx -\sin\phi\frac{\mu_0 I a^2 \sin\theta}{4r^2} \qquad (3.67)$$

$$A_y \approx \cos\phi\frac{\mu_0 I a^2 \sin\theta}{4r^2}. \qquad (3.68)$$

Hence

$$\overline{A} = \hat{x}A_x + \hat{y}A_y$$

$$= \underbrace{(-\hat{x}\sin\phi + \hat{y}\cos\phi)}_{\hat{\phi}}\frac{\mu_0 I a^2 \sin\theta}{4r^2}. \qquad (3.69)$$

Performing $\nabla \times \overline{A}$ in the spherical coordinates (r, θ, ϕ), we obtain

$$\overline{B} = \nabla \times \overline{A} = \frac{\mu_0 I a^2}{4r^3}\left(\hat{r}2\cos\theta + \hat{\theta}\sin\theta\right) \qquad (3.70)$$

whose characteristic is essentially identical with the electric dipole case (2.33). Next we introduce a magnetic dipole moment $\overline{m} = I\overline{s}$, where s $(= \pi a^2)$ is the loop area and the directions of $\overline{s}$ and I follow a right-hand rule. Figure 3.11 illustrates a magnetic dipole moment $\overline{m}$, which represents a small circular current loop located at $\overline{r}'$. The magnetic vector potential at $\overline{r}$ is compactly written as

$$\overline{A} = \frac{\mu_0\overline{m} \times \hat{R}}{4\pi R^2} \qquad (3.71)$$

where $\overline{R} = \overline{r} - \overline{r}'$. Note that expression (3.71) is a generalization of (3.69).

3.2.3 Ampère's Circuital Law

Ampère's circuital law is another important formula that relates the magnetic flux density to the steady current. We will derive Ampère's circuital law using the magnetic vector potential $\overline{A}$, although the derivation is somewhat cumbersome. Consider a current density $\overline{J}'$ of a volume v', as illustrated in Fig. 3.12. The magnetic vector potential due to $\overline{J}'$ is

$$\overline{A} = \frac{\mu_0}{4\pi}\int_{v'}\frac{\overline{J}'}{R}\,dv'. \qquad (3.72)$$

Fig. 3.11 A magnetic dipole
moment $\overline{m}$ located at $\overline{r}'$

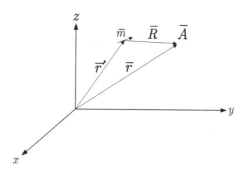

Fig. 3.12 A steady current
density flowing in v' surrounded by a closed surface s'

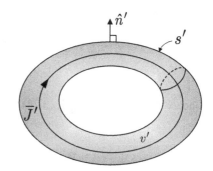

Here the integration $\int_{v'}$ is over the volume v' occupied by the current density $\overline{J}'$, which is a function of (x', y', z'). The coordinates (x', y', z') and (x, y, z) refer to the positions of source $(\overline{J}')$ and observation $(\overline{A})$, respectively. The vector $\nabla \times \overline{B}$ is written in terms of $\overline{A}$

$$\nabla \times \overline{B} = \nabla \times \nabla \times \overline{A} = \underbrace{-\nabla^2 \overline{A}}_{\mu_0 \overline{J}} + \nabla\left(\nabla \cdot \overline{A}\right) \tag{3.73}$$

where the current density $\overline{J}$ is a function of (x, y, z) and

$$\nabla \cdot \overline{A} = \nabla \cdot \left(\frac{\mu_0}{4\pi} \int_{v'} \frac{\overline{J}'}{R} dv'\right)$$

$$= \frac{\mu_0}{4\pi} \int_{v'} \nabla \cdot \left(\frac{\overline{J}'}{R}\right) dv'. \tag{3.74}$$

In order to evaluate $\nabla \cdot \overline{A}$ for the steady current $\left(\nabla' \cdot \overline{J}' = 0\right)$, we consider the following two divergences:

$$\nabla \cdot \left(\frac{\overline{J'}}{R} \right) = \overline{J'} \cdot \nabla \left(\frac{1}{R} \right) + \frac{1}{R} \underbrace{\nabla \cdot \overline{J'}}_{0} \tag{3.75}$$

$$\nabla' \cdot \left(\frac{\overline{J'}}{R} \right) = \overline{J'} \cdot \nabla' \left(\frac{1}{R} \right) + \frac{1}{R} \underbrace{\nabla' \cdot \overline{J'}}_{0} \tag{3.76}$$

where

$$R = \sqrt{(x - x')^2 + (y - y')^2 + (z - z')^2} \tag{3.77}$$

$$\nabla = \hat{x} \frac{\partial}{\partial x} + \hat{y} \frac{\partial}{\partial y} + \hat{z} \frac{\partial}{\partial z} \tag{3.78}$$

$$\nabla' = \hat{x} \frac{\partial}{\partial x'} + \hat{y} \frac{\partial}{\partial y'} + \hat{z} \frac{\partial}{\partial z'} . \tag{3.79}$$

Since $\nabla \left(\frac{1}{R} \right) = -\nabla' \left(\frac{1}{R} \right)$, (3.74) becomes

$$\nabla \cdot \overline{A} = -\frac{\mu_0}{4\pi} \int_{v'} \nabla' \cdot \left(\frac{\overline{J'}}{R} \right) dv'. \tag{3.80}$$

Since the operations $\int_{v'}$ and ∇' both are with respect to (x', y', z'), it is possible to apply the divergence theorem to (3.80). The result is

$$\nabla \cdot \overline{A} = -\frac{\mu_0}{4\pi} \oint_{s'} \left(\frac{\overline{J'}}{R} \right) \cdot d\overline{s'}$$

$$= -\frac{\mu_0}{4\pi} \oint_{s'} \frac{\overline{J'} \cdot \hat{n'}}{R} ds' \tag{3.81}$$

where $\hat{n'}$ is a unit vector normal to the surface s'. Note that $\overline{J'}$ is the steady current density, which flows along the loop and closes upon itself. Since the surface s' encloses $\overline{J'}$, no current flows through s' such as $\overline{J'} \cdot \hat{n'} = 0$ on s'. Therefore $\nabla \cdot \overline{A} = 0$ and (3.73) becomes

$$\nabla \times \overline{B} = \mu_0 \overline{J}. \tag{3.82}$$

Ampère's circuital law

Fig. 3.13 A current density $\overline{J}$ passing through a surface s

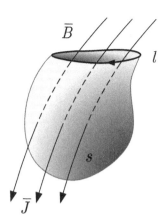

Next we will transform Ampère's circuital law into an integral form. Consider a current $\overline{J}$ that produces $\overline{B}$ in Fig. 3.13. Applying Stokes's theorem

$$\int_s \nabla \times \overline{B} \cdot d\overline{s} = \oint_l \overline{B} \cdot d\overline{r} \tag{3.83}$$

to Ampère's circuital law, we get

$$\oint_l \overline{B} \cdot d\overline{r} = \mu_0 \underbrace{\int_s \overline{J} \cdot d\overline{s}}_{I}. \tag{3.84}$$

Ampère's circuital law in integral form

Note that I is the total current passing through a surface s, where the directions of l and $\overline{s}$ follow a right-hand rule. Ampère's circuital law states that the line integration of $\overline{B}$ along the closed path l $\left(\oint_l \overline{B} \cdot d\overline{r} \right)$ equals $\mu_0 I$, where I is the current enclosed by l.

3.3 Magnetic Forces and Torques

Current-carrying wires placed in magnetic fields experience magnetic forces. It is of practical importance to further investigate these phenomena since electric motors have direct relevance to magnetic forces. We consider a loop l carrying a filamentary

Fig. 3.14 A current loop l
where $\bar{r} = \hat{x}x + \hat{y}y + \hat{z}z$

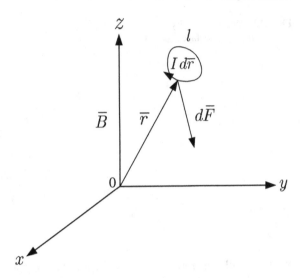

current, as shown in Fig. 3.14. A current loop l placed in a magnetic flux density $\overline{B}$
experiences a force $\overline{F}$. Suppose that a differential force $d\overline{F}$ is exerted on a differential body $I d\bar{r}$. This force can be determined by Ampère's law of force, where the
differential force acting on $d\bar{r}$ is

$$d\overline{F} = I d\bar{r} \times \overline{B}. \tag{3.85}$$

For simplicity we assume that the magnetic flux density is uniform such as

$$\overline{B} = \hat{x} B_x + \hat{y} B_y + \hat{z} B_z \tag{3.86}$$

where B_x, B_y, and B_z are constants. If $\overline{B}$ is uniform, the total force on the loop is
zero since

$$\overline{F} = I \underbrace{\left(\oint_l d\bar{r} \right)}_{0} \times \overline{B} = 0 \tag{3.87}$$

where the symbol $\oint_l$ denotes a line integral around the closed loop l. Next we will
evaluate the torque, which provides an additional information about a rotational
motion of the current loop. We first introduce the definition of torque. Suppose
that a differential force $d\overline{F}$ acts on a differential loop $d\bar{r}$ having a position vector
$\bar{r}$, as shown in Fig. 3.14. Then the differential torque acting on $d\bar{r}$ with respect to
the origin is

$$d\overline{\tau} = \overline{r} \times d\overline{F}.$$ (3.88)

Hence the total torque on the loop l is

$$\overline{\tau} = \oint_l \overline{r} \times d\overline{F}$$

$$= \oint_l \overline{r} \times \left(I d\overline{r} \times \overline{B} \right)$$

$$= \underbrace{\oint_l I d\overline{r} \left(\overline{r} \cdot \overline{B} \right)}_{I_1} - \underbrace{\oint_l \overline{B} \left(\overline{r} \cdot I d\overline{r} \right)}_{I_2}.$$ (3.89)

We evaluate I_2 and I_1 separately as

$$I_2 = \overline{B} I \oint_l \overline{r} \cdot d\overline{r} = \overline{B} I \int_s \underbrace{\left(\nabla \times \overline{r} \right) \cdot d\overline{s}}_{0} = 0.$$ (3.90)

We have used Stokes's theorem in (3.90). Furthermore we have

$$I_1 = I \oint_l d\overline{r} \left(\overline{r} \cdot \overline{B} \right)$$

$$= I \int_s d\overline{s} \times \underbrace{\left[\nabla \left(\overline{r} \cdot \overline{B} \right) \right]}_{\overline{B}}$$

$$= I \left(\int_s d\overline{s} \right) \times \overline{B}$$

$$= I \overline{s} \times \overline{B}.$$ (3.91)

We have used in (3.91) an alternative form of Stokes's theorem

$$\oint_l \phi \, d\overline{r} = \int_s d\overline{s} \times \nabla \phi.$$ (3.92)

Equation (3.92) can be derived by substituting $\overline{A} = \overline{a}\phi$ into Stokes's theorem $\left(\int_s \nabla \times \overline{A} \cdot d\overline{s} = \oint_l \overline{A} \cdot d\overline{r} \right)$, where $\overline{a}$ is a vector having a constant magnitude and a constant direction. Note that s is any open surface whose rim is l. Substituting (3.90) and (3.91) into (3.89), we obtain the torque expression

Fig. 3.15 Planar rectangular
loop in a magnetic field

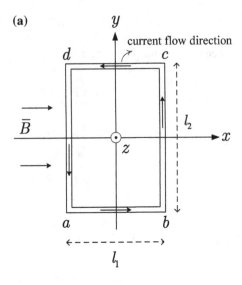

(a)

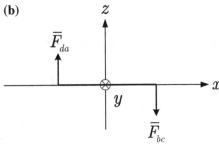

(b)

$$\overline{\tau} = \overline{m} \times \overline{B} \qquad\qquad (3.93)$$

torque on the loop

where $\overline{m}$ $(= I\overline{s})$ is the magnetic dipole moment. When a current loop is placed in
a uniform magnetic field, a torque is generated on a current loop. This principle is a
basic mechanism of DC motors that convert electric energy into mechanical energy.
The following example will further elucidate this principle.

Example 3.5 Torque on a rectangular loop.

Figure 3.15a shows a rectangular loop ($l_1 \times l_2$) carrying a current I placed on the
x-y plane. The center of the loop is located at the origin. When a uniform magnetic
field $\overline{B} = \hat{x}B$ is applied, evaluate the torque exerted on the loop.

Solution:

1. **Approach using $\overline{\tau} = \overline{m} \times \overline{B}$:**
   The total torque acting on the loop is

   $$\overline{\tau} = \overline{m} \times \overline{B}$$

   $$= I\overline{s} \times \overline{B}$$

   $$= I\hat{z}l_1l_2 \times \hat{x}B$$

   $$= \hat{y}Il_1l_2B \qquad (3.94)$$

   which indicates that the loop tends to rotate about the y-axis.

2. **Approach using $d\overline{\tau} = \overline{r} \times d\overline{F}$:**
   It is possible to evaluate $\overline{\tau}$ without recourse to the relation $\overline{\tau} = \overline{m} \times \overline{B}$. The approach using $\overline{r} \times d\overline{F}$ will give us an intuitive feel as to what the torque does to the loop. The individual forces acting on the sides (ab, bc, cd, and da) are

   $$\overline{F}_{ab} = I\underbrace{\left(\int_{ab} d\overline{r}\right)}_{\hat{x}l_1} \times \overline{B} = 0 \qquad (3.95)$$

   $$\overline{F}_{bc} = I\underbrace{\left(\int_{bc} d\overline{r}\right)}_{\hat{y}l_2} \times \overline{B} = -\hat{z}Il_2B \qquad (3.96)$$

   $$\overline{F}_{cd} = I\underbrace{\left(\int_{cd} d\overline{r}\right)}_{-\hat{x}l_1} \times \overline{B} = 0 \qquad (3.97)$$

   $$\overline{F}_{da} = I\underbrace{\left(\int_{da} d\overline{r}\right)}_{-\hat{y}l_2} \times \overline{B} = \hat{z}Il_2B. \qquad (3.98)$$

As expected, the total force is shown to be zero by the relation

$$\overline{F} = I\oint_l d\overline{r} \times \overline{B} = \overline{F}_{ab} + \overline{F}_{bc} + \overline{F}_{cd} + \overline{F}_{da} = 0. \qquad (3.99)$$

The equal but opposite forces $\overline{F}_{bc}$ and $\overline{F}_{da}$, exerted on the sides bc and da, attempt to rotate the loop around the y-axis, as illustrated in Fig. 3.15b. This mechanical rotation can be explained in terms of a torque. The torque on the side bc is

$$\overline{T}_{bc} = \int_{bc} \overline{r} \times d\overline{F}$$

$$= \int_{y=-l_2/2}^{l_2/2} \underbrace{(\hat{x}l_1/2 + \hat{y}y)}_{\overline{r}} \times \underbrace{BI(\hat{y}\,dy \times \hat{x})}_{d\overline{F}}$$

$$= \hat{y}BIl_1l_2/2. \tag{3.100}$$

Similarly the torques on the remaining sides are

$$\overline{T}_{da} = \int_{da} \overline{r} \times d\overline{F}$$

$$= \int_{y=l_2/2}^{-l_2/2} \underbrace{(-\hat{x}l_1/2 + \hat{y}y)}_{\overline{r}} \times \underbrace{BI(\hat{y}\,dy \times \hat{x})}_{d\overline{F}}$$

$$= \hat{y}BIl_1l_2/2 \tag{3.101}$$

$$\overline{T}_{ab} = \overline{T}_{cd} = 0. \tag{3.102}$$

Hence the total torque is

$$\overline{T} = \overline{T}_{ab} + \overline{T}_{bc} + \overline{T}_{cd} + \overline{T}_{da} = \hat{y}Il_1l_2B. \tag{3.103}$$

3.4 Magnetic Materials and Boundary Conditions

3.4.1 Magnetic Materials

Any materials are composed of electrons and nuclei. Electrons orbiting positive nuclei in material media are regarded as infinitesimal current loops with magnetic dipole moments. Discussions of magnetic dipole moments are provided in Sect. 3.2.2. If the external magnetic flux density is absent, infinitesimal current loops are randomly oriented with zero net magnetic dipole moment. However, when an external magnetic flux density is applied to the material, orbiting electrons are realigned, and the net magnetic dipole moment becomes nonzero. The realigned orbiting electrons are considered an induced current density that produces the net magnetic dipole moment. Consider a differential volume dv' whose induced magnetic dipole moment per unit volume is given by a magnetization $\overline{M}'$ $(= \overline{M}(\overline{r}'))$. The differential magnetic vector potential at (x, y, z) due to dv' at (x', y', z') is then

$$d\overline{A} = \frac{\mu_0 \overline{M}' \times \hat{R}}{4\pi R^2} dv' \tag{3.104}$$

which has relevance to the magnetic dipole in Example 3.4. Here

$$R = \sqrt{(x - x')^2 + (y - y')^2 + (z - z')^2}. \tag{3.105}$$

The total magnetic vector potential at (x, y, z) due to v' is

$$\overline{A} = \int_{v'} \frac{\mu_0 \overline{M}' \times \hat{R}}{4\pi R^2} dv'. \tag{3.106}$$

Note that

$$\nabla'\left(\frac{1}{R}\right) = \frac{\hat{R}}{R^2} \tag{3.107}$$

where ∇' is the differential operator *del* with respect to the primed variables (x', y', z'). Therefore the magnetic vector potential is rewritten as

$$\overline{A} = \frac{\mu_0}{4\pi} \int_{v'} \overline{M}' \times \nabla'\left(\frac{1}{R}\right) dv'. \tag{3.108}$$

Furthermore we note the vector identity

$$\nabla' \times \left(\frac{\overline{M}'}{R}\right) = -\overline{M}' \times \nabla'\left(\frac{1}{R}\right) + \frac{\nabla' \times \overline{M}'}{R}. \tag{3.109}$$

Hence

$$\overline{A} = \frac{\mu_0}{4\pi}\left[\int_{v'} -\nabla' \times \left(\frac{\overline{M}'}{R}\right) dv' + \int_{v'} \frac{\nabla' \times \overline{M}'}{R} dv'\right]$$

$$= \frac{\mu_0}{4\pi}\left[\oint_{s'} \frac{1}{R}\underbrace{\left(\overline{M}' \times \hat{n}'\right)}_{\overline{J}'_{ms}} ds' + \int_{v'} \frac{1}{R}\underbrace{\left(\nabla' \times \overline{M}'\right)}_{\overline{J}'_{mv}} dv'\right] \tag{3.110}$$

where $\hat{n}'$ is a unit vector normal to the surface ds' and away from the volume v'. We have used in (3.110) an alternative form of divergence theorem

$$\int_{v'} \nabla' \times \overline{C} \, dv' = -\oint_{s'} \overline{C} \times \hat{n}' \, ds'. \tag{3.111}$$

Equation (3.111) can be derived by substituting $\overline{U} = \overline{a} \times \overline{C}$ into the divergence theorem

$$\int_{v'} \nabla' \cdot \overline{U} \, dv' = \oint_{s'} \overline{U} \cdot \hat{n}' \, ds' \tag{3.112}$$

where $\overline{a}$ is a vector having a constant magnitude and a constant direction. We recall that the magnetic vector potential due to the volume current density $\overline{J}'$ is defined as

$$\overline{A} = \frac{\mu_0}{4\pi} \int_{v'} \frac{\overline{J}'}{R} \, dv'. \tag{3.113}$$

In view of (3.113), the terms $\overline{J}'_{mv}$ and $\overline{J}'_{ms}$ in (3.110) are regarded as a bound volume current density and a bound surface current density, respectively. The magnetic vector potential due to a magnetic material can be obtained by integrating contributions from $\overline{J}'_{mv}$ and $\overline{J}'_{ms}$, which are located inside the material volume and at the boundary surface, respectively. Therefore an additional contribution $\overline{M}$ is generated inside the material as

$$\nabla \times \overline{M} = \overline{J}_{mv}. \tag{3.114}$$

To account for this additional contribution to the magnetic flux density $\overline{B}$, we rewrite Ampère's circuital law for a free volume current density $\overline{J}$ and a bound volume current density $\overline{J}_{mv}$ as

$$\nabla \times \overline{B} = \mu_0 \left(\overline{J} + \overline{J}_{mv} \right). \tag{3.115}$$

Expressions (3.115) and (3.114) yield

$$\nabla \times \underbrace{\left(\frac{\overline{B}}{\mu_0} - \overline{M} \right)}_{\overline{H}} = \overline{J} \tag{3.116}$$

where $\overline{H}$ is defined as a magnetic field intensity. Assuming that a magnetic material is linear such as $\overline{M} = \chi_m \overline{H}$, we write the magnetic flux density as

$$\overline{B} = \mu_0 \underbrace{(1 + \chi_m)}_{\mu_r} \overline{H} = \mu \overline{H} \tag{3.117}$$

where μ, χ_m, and μ_r are the permeability, the magnetic susceptibility, and the relative permeability of material medium, respectively. If the magnetic material is linear, isotropic, and homogeneous, then χ_m is constant. The permeability of most simple materials is μ_0. Ampère's circuital law for material media should be rewritten as

$$\nabla \times \overline{H} = \overline{J} \qquad (3.118)$$

Ampère's circuital law for material media

where $\overline{J}$ is a free volume current density. Ampère's circuital law can be written in integral form as well. Applying Stokes's theorem to Ampère's circuital law yields

$$\oint_l \overline{H} \cdot d\overline{r} = \underbrace{\int_s \overline{J} \cdot d\overline{s}}_{I}. \qquad (3.119)$$

Ampère's circuital law in integral form

Ampère's circuital law states that the line integration of $\overline{H}$ along the closed path l $\left(\oint_l \overline{H} \cdot d\overline{r} \right)$ equals the current I enclosed by l.

3.4.2 Boundary Conditions

A magnetic field intensity $\overline{H}$ and a magnetic flux density $\overline{B}$ across the boundary follow the magnetic Gauss's law and Ampère's circuital law. Boundary conditions between different magnetic media can be derived from the magnetic Gauss's law and Ampère's circuital law. Figure 3.16 shows a small pill box that is situated across

Fig. 3.16 Magnetic field across the boundary with a small pill box

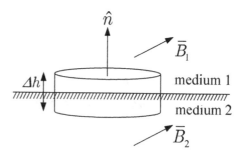

medium 1 and medium 2. Applying the magnetic Gauss's law $\oint_s \overline{B} \cdot d\overline{s} = 0$ to a
small pill box and shrinking $\Delta h \to 0$, we obtain the boundary condition

$$\hat{n} \cdot (\overline{B}_1 - \overline{B}_2) = 0 \qquad\qquad (3.120)$$

normal boundary condition

where $\hat{n}$ denotes a normal unit vector pointing away from medium 2. The above
expression states that the normal component of $\overline{B}$ must be continuous across the
boundary. Figure 3.17 shows a closed path encircling the boundary between medium
1 and medium 2. Here we introduce three orthogonal vectors $(\hat{u}, \hat{t}, \hat{n})$, where the
unit vectors $\hat{u}$ and $\hat{t}$ constitute a tangential plane and $\hat{n}$ is a unit vector normal to the
tangential plane. Applying Ampère's circuital law

$$\oint_l \overline{H} \cdot d\overline{r} = \Delta I \qquad\qquad (3.121)$$

to a small rectangle in Fig. 3.17 and shrinking $\Delta h \to 0$, we obtain

$$\underbrace{\overline{H}_2 \cdot \hat{t}}_{H_{2t}} \Delta l - \underbrace{\overline{H}_1 \cdot \hat{t}}_{H_{1t}} \Delta l = \Delta I \qquad\qquad (3.122)$$

where ΔI is a current that flows in a direction $(\hat{u})$ normal to the rectangular area at
the boundary between medium 1 and medium 2. Equation (3.122) is rewritten as

$$H_{2t} - H_{1t} = J_u. \qquad\qquad (3.123)$$

This condition states that the tangential magnetic fields (H_{2t} and H_{1t}) are discontinuous by the amount J_u ($= \Delta I / \Delta l$), which is a surface current density flowing in

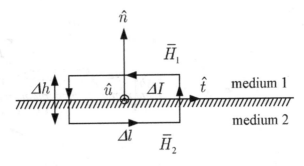

Fig. 3.17 Magnetic field across the boundary with a small rectangle

the $\hat{u}$-direction. Similarly, by applying Ampère's circuital law to a small rectangle on the $\hat{n}$-$\hat{u}$ plane, we obtain

$$H_{2u} - H_{1u} = -J_t \tag{3.124}$$

where H_{2u} and H_{1u} are tangential components in the $\hat{u}$-direction and J_t is a surface current density flowing in the $\hat{t}$-direction. We will rewrite (3.123) and (3.124) more compactly by using a unit normal vector. Note

$$\hat{n} \times \left(\overline{H}_1 - \overline{H}_2 \right) = \underbrace{(H_{2t} - H_{1t})}_{J_u} \hat{u} + \underbrace{(H_{1u} - H_{2u})}_{J_t} \hat{t}. \tag{3.125}$$

Since the total surface current density is $\overline{J}_s = \hat{t} J_t + \hat{u} J_u$, we obtain the tangential boundary condition

$$\hat{n} \times \left(\overline{H}_1 - \overline{H}_2 \right) = \overline{J}_s. \tag{3.126}$$

tangential boundary condition

This boundary condition states that the tangential magnetic field is discontinuous by the amount $\overline{J}_s$. When the boundary is devoid of $\overline{J}_s$, the tangential components of $\overline{H}$ should be continuous across the boundary.

Example 3.6 Magnetic field due to a current-carrying cylinder.
In Fig. 3.18, an infinitely long circular cylinder of radius a carries a current I, which is uniformly distributed over the cross section. Determine the magnetic field $\overline{H}$.

Solution: Figure 3.19 illustrates a cross section of current-carrying cylinder. For two identical currents i symmetric with respect to ρ, the resultant magnetic field $\overline{H}_p$ at position P points in the ϕ-direction. Therefore the total magnetic field due to I has a ϕ component H_ϕ. The field H_ϕ is a function of ρ.

Fig. 3.18 A long circular cylinder carrying a current I

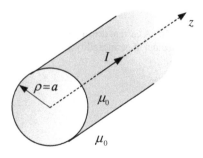

Fig. 3.19 Two identical
currents i with respect to ρ

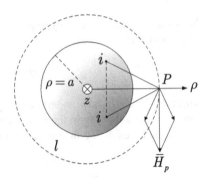

1. When $\rho \leq a$, we choose a circular path l that encircles the current $I\left(\dfrac{\rho}{a}\right)^2$.
   Applying Ampère's circuital law gives

$$\oint_l H_\phi \, dr = I\left(\frac{\rho}{a}\right)^2 \implies H_\phi = \frac{I\rho}{2\pi a^2}. \tag{3.127}$$

2. When $\rho \geq a$, we choose a circular path l that encircles I. Applying Ampère's
   circuital law gives

$$\oint_l H_\phi \, dr = I \implies H_\phi = \frac{I}{2\pi\rho}. \tag{3.128}$$

Note that the tangential magnetic field H_ϕ is continuous across $\rho = a$. The magnetic
field decays as $\dfrac{1}{\rho}$ outside the current-carrying cylinder.

3.5 Poisson's Equation for Magnetic Fields

Figure 3.20 shows a steady current density $\overline{J}$ placed in a homogeneous medium
of permeability μ. The current density $\overline{J}$ and the magnetic field $\overline{H}$ are related by

Fig. 3.20 A current density $\overline{J}$
in a medium of permeability μ

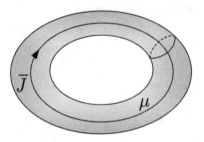

Ampère's circuital law $\nabla \times \overline{H} = \overline{J}$. The magnetic flux density $\overline{B}$ is given by the magnetic vector potential $\overline{A}$ as

$$\overline{B} = \nabla \times \overline{A} \Longrightarrow \overline{H} = \frac{1}{\mu}\nabla \times \overline{A}. \qquad (3.129)$$

We will show that Poisson's equation governs $\overline{A}$ in a medium of μ. Substituting (3.129) into Ampère's circuital law, we obtain

$$\nabla \times \nabla \times \overline{A} = \mu \overline{J} \qquad (3.130)$$

which is rewritten as

$$-\nabla^2 \overline{A} + \nabla\left(\nabla \cdot \overline{A}\right) = \mu\overline{J}. \qquad (3.131)$$

Since $\nabla \cdot \overline{A} = 0$, we obtain

$$\nabla^2 \overline{A} = -\mu \overline{J}. \qquad (3.132)$$

Poisson's equation for $\overline{A}$

In particular, when a region of interest is devoid of currents ($\overline{J} = 0$), Poisson's equation is reduced to

$$\nabla^2 \overline{A} = 0. \qquad (3.133)$$

Generally, it is possible to find the magnetic fields within bounded areas by solving Poisson's equation subject to the boundary conditions. In the following example, we will solve Poisson's equation to determine the magnetic field resulting from a long current.

Example 3.7 Field calculation using Poisson's equation.
Consider an infinitely long circular cylinder of radius a carrying a uniform current I, as shown in Fig. 3.18. Determine the magnetic field $\overline{H}$ by using Poisson's equation.

Solution: Poisson's equation for $\overline{A}$ is

$$\nabla^2 \overline{A} = \begin{cases} -\hat{z}\mu_0 J & \text{for} \quad \rho \le a \\ 0 & \text{for} \quad \rho > a \end{cases} \qquad (3.134)$$

where $J = \dfrac{I}{\pi a^2}$. Since the current density has a z-component, the magnetic vector potential $\overline{A}$ is assumed to have a z-component (A_z) only. Poisson's equation is rewritten as

$$\nabla^2 A_z = \begin{cases} -\mu_0 J & \text{for} \quad \rho \le a \\ 0 & \text{for} \quad \rho > a. \end{cases} \tag{3.135}$$

1. First consider the region $\rho \le a$. Since A_z has no variation in z and ϕ, (3.135) is rewritten as

$$\frac{1}{\rho}\frac{d}{d\rho}\left(\rho\frac{dA_z}{d\rho}\right) = -\mu_0 J. \tag{3.136}$$

This yields

$$\frac{dA_z}{d\rho} = -\frac{\mu_0 J \rho}{2} + \frac{c_1}{\rho} \tag{3.137}$$

where c_1 is an unknown coefficient to be determined. From $\overline{B} = \nabla \times \overline{A}$, we obtain the magnetic flux density within the cylinder as

$$B_\phi^1 = -\frac{dA_z}{d\rho} = \frac{\mu_0 J \rho}{2} - \frac{c_1}{\rho}. \tag{3.138}$$

Note that the magnetic flux density has a component in the ϕ-direction. Since B_ϕ^1 must be finite at $\rho = 0$, $c_1 = 0$.

2. Next consider the region $\rho > a$. The governing equation is

$$\frac{1}{\rho}\frac{d}{d\rho}\left(\rho\frac{dA_z}{d\rho}\right) = 0. \tag{3.139}$$

This yields

$$\frac{dA_z}{d\rho} = \frac{c_2}{\rho}. \tag{3.140}$$

The magnetic flux density outside the cylinder is

$$B_\phi^0 = -\frac{dA_z}{d\rho} = -\frac{c_2}{\rho}. \tag{3.141}$$

The magnetic field H_ϕ is continuous at $\rho = a$ as

$$\left.\frac{B_\phi^1}{\mu_0}\right|_{\rho=a} = \left.\frac{B_\phi^0}{\mu_0}\right|_{\rho=a} \tag{3.142}$$

which gives

$$c_2 = -\frac{Ja^2\mu_0}{2} = -\frac{\mu_0 I}{2\pi}.$$ (3.143)

Therefore the magnetic field is

$$H_\phi = \begin{cases} \dfrac{I\rho}{2\pi a^2} & \text{for } \rho \leq a \\[2mm] \dfrac{I}{2\pi\rho} & \text{for } \rho > a. \end{cases}$$ (3.144)

3.6 Problems for Chapter 3

1. Figure 3.21 shows the cross section of a coaxial cable consisting of two concentric conducting cylinders of length l with radii $\rho = a$ and $\rho = b$. The interior region $(a < \rho < b)$ is filled with a lossy dielectric of conductivity σ. When a DC voltage is applied between two concentric conducting cylinders, a leakage current flows in the radial (ρ) direction. Determine the leakage resistance of this lossy dielectric. **Hint**: Use Laplace's equation to determine the voltage within the lossy medium and apply Ohm's law to determine the current.

2. In Fig. 3.22 a circular loop with a radius a carrying a current I lies on the x-y plane. Find the magnetic field at the center $(x = y = z = 0)$ of the circular loop. **Hint**: Use the Biot-Savart law to determine a magnetic flux density from a differential current.

3. Figure 3.23 shows a thin circular disk of radius a carrying a uniform surface charge density ρ_s centered on the x-y plane. When the disk rotates about the z-axis with an angular speed ω, evaluate the magnetic flux density on the positive z-axis.

Fig. 3.21 A cross section of coaxial cable

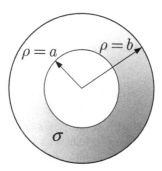

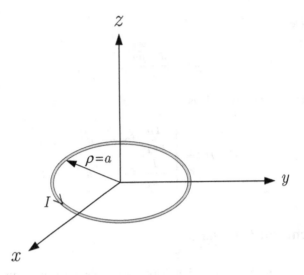

Fig. 3.22 A circular current loop with a radius a

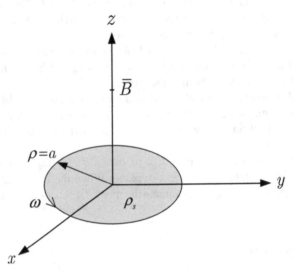

Fig. 3.23 A circular charged disk with a radius a rotating with an angular speed ω

4. Figure 3.24 illustrates an infinite sheet of the uniform surface current density $\overline{J}_s = \hat{y} J_s$ placed at $z = 0$. Find the magnetic field.
   Hint: Use Ampère's circuital law.
5. Figure 3.25 shows a long thin current strip of width w carrying a current I. An equal current-carrying wire is placed at $z = h$ in parallel with the strip. Calculate the magnetic force per unit length exerted on the strip.
   Hint: Use Ampère's law of force.

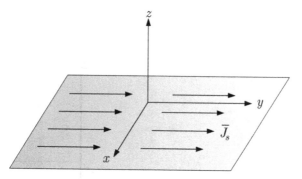

Fig. 3.24 An infinite current sheet $\bar{J}_s$

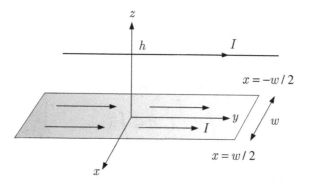

Fig. 3.25 A current-carrying strip and a current-carrying wire

6. Figure 3.26 shows an infinitely long line current I in free space. Evaluate the magnetic field by using the following three different approaches:

- Biot-Savart law

$$\bar{B} = \frac{\mu_0}{4\pi} \oint_{l'} \frac{I' d\bar{r}' \times \hat{R}}{R^2} \qquad (3.145)$$

- magnetic vector potential expression

$$\bar{A} = \frac{\mu_0}{4\pi} \int_{v'} \frac{\bar{J}'}{R} \, dv' \qquad (3.146)$$

- Ampère's circuital law

$$\oint_l \bar{B} \cdot d\bar{r} = \mu_0 I. \qquad (3.147)$$

Fig. 3.26 An infinitely long
line current

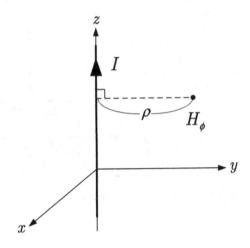

7. Derive (3.92) by substituting $\overline{A} = \overline{a}\phi$ ($\overline{a}$: a vector having a constant magnitude and a constant direction) into Stokes's theorem

$$\int_s \nabla \times \overline{A} \cdot d\overline{s} = \oint_l \overline{A} \cdot d\overline{r}. \qquad (3.148)$$

Hint: Use the identity $(\nabla\phi \times \overline{a}) \cdot d\overline{s} = \overline{a} \cdot (d\overline{s} \times \nabla\phi)$.

8. The magnetic vector potential $\overline{A}$ at $r \gg a$ for a magnetic dipole is given by (3.69) as

$$\overline{A} = \hat{\phi}\frac{\mu_0 I a^2 \sin\theta}{4r^2}. \qquad (3.149)$$

Show that $\overline{A}$ satisfies $\nabla^2 \overline{A} = 0$.
Hint: Use the identity $\nabla^2 \overline{A} = \nabla \left(\nabla \cdot \overline{A}\right) - \nabla \times \nabla \times \overline{A}$ in spherical coordinates.

Chapter 4
Faraday's Law of Induction

4.1 Faraday's Law

We will first introduce Faraday's law (Fig. 4.1) by using a stationary wire loop in Fig. 4.2. When a closed loop of wire is in a time-varying magnetic flux density, a time-varying electric field is induced in the wire loop. The induced electric field exerts a force on free charges of the wire loop, thus producing a current on the loop. The induced electric field is often interpreted in terms of an induced electromotive force (emf). This principle of electromagnetic induction, discovered experimentally by Michael Faraday, can be written as follows: when a loop is in a time-varying magnetic flux Φ_m, as shown in Fig. 4.2, the emf V induced around the loop is

$$V = -\frac{d\Phi_m}{dt} \qquad (4.1)$$

Faraday's law

where $\dfrac{d}{dt}$ is differentiation with respect to time t. Faraday's law states that the emf induced on a closed loop equals the negative time rate of the magnetic flux encircled by the loop. The negative sign indicates that the induced emf produces a current whose magnetic flux density suppresses a change in the original magnetic flux. This phenomenon is referred to as Lenz's law. Here the emf induced on a closed loop l and the magnetic flux across a surface s are given by

$$V = \oint_l \bar{\mathcal{E}} \cdot d\bar{r} \qquad (4.2)$$

$$\Phi_m = \int_s \bar{B} \cdot d\bar{s} \qquad (4.3)$$

H. J. Eom, *Primary Theory of Electromagnetics*, Power Systems,
DOI: 10.1007/978-94-007-7143-7_4, © Springer Science+Business Media Dordrecht 2013

Fig. 4.1 Faraday's law in differential form

$$\nabla \times \bar{\mathcal{E}} = -\frac{\partial \bar{B}}{\partial t}$$

Faraday's law

Fig. 4.2 A stationary loop in a time-varying magnetic flux density $\bar{B}$

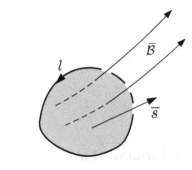

Fig. 4.3 An area $\bar{s}$ encircled by a closed loop l

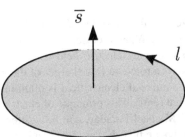

where l is a loop encircling a loop area s. Apparently, there exist sign ambiguities associated with l and $\bar{s}$. How do we choose the directions of l and $\bar{s}$? This choice is at our disposal as long as the directions of l and $\bar{s}$ follow a right-hand rule, as depicted in Fig. 4.3. The contour shape of $\bar{s}$ may be arbitrary. Faraday's law in integral form is rewritten as

$$\oint_l \bar{\mathcal{E}} \cdot d\bar{r} = -\frac{d}{dt} \int_s \bar{B} \cdot d\bar{s}. \qquad (4.4)$$

Faraday's law in integral form

Faraday's law, although initially discovered with a conducting loop, is applicable even to a nonconducting loop. In other words, there is no restriction on the type of loop and Faraday's law is valid for any hypothetical closed loops.

A loop with a gap: We will consider a perfectly conducting wire with an infinitesimal gap, as shown in Fig. 4.4. The induced emf $\mathcal{V}$ exerts a force on free charges on a

Fig. 4.4 A loop with an infinitesimal gap in a time-varying
magnetic flux density $\overline{B}$

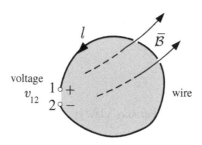

conducting wire and these charges accumulate at terminals 1 and 2. If fields are slowly varying, a voltage, which is a potential difference in electrostatics, is acceptable for our discussion. A charge accumulation at terminals 1 and 2 is said to produce a voltage v_{12} equal to $\mathcal{V}$. The proof is as follows:

Proof The induced emf $\mathcal{V}$ is

$$\mathcal{V} = \oint_l \overline{\mathcal{E}} \cdot d\overline{r}$$

$$= \int_{wire} \overline{\mathcal{E}} \cdot d\overline{r} + \int_1^2 \overline{\mathcal{E}} \cdot d\overline{r} \tag{4.5}$$

where the symbol $\displaystyle\int_{wire}$ denotes a line integral along a perfectly conducting wire. The tangential electric field on the perfectly conducting wire should be zero, leading to the relation $\displaystyle\int_{wire} \overline{\mathcal{E}} \cdot d\overline{r} = 0$. Note

$$\int_1^2 \overline{\mathcal{E}} \cdot d\overline{r} = \underbrace{-\int_2^1 \overline{\mathcal{E}} \cdot d\overline{r}}_{v_{12}}. \tag{4.6}$$

Hence the induced emf $\mathcal{V}$ is equal to the voltage v_{12} ($\mathcal{V} = v_{12}$).

Faraday's law in differential form: At this point, it is instructive to convert Faraday's law into a differential form. Let us first consider Stokes's theorem

$$\oint_l \overline{\mathcal{E}} \cdot d\overline{r} = \int_s \nabla \times \overline{\mathcal{E}} \cdot d\overline{s}. \tag{4.7}$$

Therefore Faraday's law is rewritten as

$$\int_s \nabla \times \overline{\mathcal{E}} \cdot d\overline{s} = -\frac{d}{dt} \int_s \overline{B} \cdot d\overline{s}. \tag{4.8}$$

In the case of a stationary loop considered in Fig. 4.2, the area s is time-invariant, hence

$$\frac{d}{dt}\int_s \overline{B}\cdot d\overline{s} = \int_s \frac{\partial \overline{B}}{\partial t}\cdot d\overline{s}. \qquad (4.9)$$

Finally, Faraday's law is

$$\nabla \times \overline{\mathcal{E}} = -\frac{\partial \overline{B}}{\partial t}. \qquad (4.10)$$

Faraday's law in differential form

Faraday's law in differential form is considered one of the fundamental laws in electromagnetic theory. It should be emphasized that Faraday's law in differential form is valid for any time-varying electromagnetic fields. For instance, although Faraday's law in differential form was derived by using a stationary conducting loop, it applies even to a moving loop. A moving loop will be further discussed next.

A moving loop: This particular case becomes important in understanding the principle of rotating electric generators. Figure 4.5 shows a loop that moves with a velocity $\overline{u}$ in $\overline{B}$. The loop does not have to retain a fixed shape in general. We consider a charge q on the moving loop. It is convenient to introduce a frame of coordinates moving with a velocity $\overline{u}$. The moving charge q is then considered stationary to the observer in the moving frame of coordinates. According to Lorentz's force, the force exerted on q in the moving frame of coordinates is $\overline{\mathcal{F}}' = q\overline{\mathcal{E}}'$ while the force exerted on q in the laboratory frame of coordinates is $\overline{\mathcal{F}} = q\left(\overline{\mathcal{E}} + \overline{u} \times \overline{B}\right)$. Here $\overline{\mathcal{E}}'$ is the electric field in the frame of coordinates moving with a velocity $\overline{u}$, whereas $\overline{\mathcal{E}}$ is the electric field in the laboratory frame of coordinates. Since $\overline{\mathcal{F}}' = \overline{\mathcal{F}}$, we obtain the relation

$$\overline{\mathcal{E}}' = \overline{\mathcal{E}} + \overline{u} \times \overline{B}. \qquad (4.11)$$

Hence Faraday's law in integral form is rewritten as

Fig. 4.5 A loop moving with a velocity $\overline{u}$

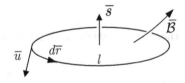

$$\oint_l \overline{\mathcal{E}}' \cdot d\overline{r} = -\frac{d}{dt} \int_s \overline{B} \cdot d\overline{s} \tag{4.12}$$

$$\underbrace{\phantom{\oint_l \overline{\mathcal{E}}' \cdot d\overline{r}}}_{\mathcal{V}}$$

where $\mathcal{V}$ is the induced emf. Equation (4.12) represents Faraday's law in integral form that is valid for a moving loop. Next we will derive different integral forms of Faraday's law starting from Faraday's law in differential form

$$\nabla \times \overline{\mathcal{E}} = -\frac{\partial \overline{B}}{\partial t}. \tag{4.13}$$

Integrating this over $\overline{s}$ and applying Stokes's theorem, we obtain

$$\oint_l \overline{\mathcal{E}} \cdot d\overline{r} = -\int_s \frac{\partial \overline{B}}{\partial t} \cdot d\overline{s}. \tag{4.14}$$

Furthermore, substituting $\overline{\mathcal{E}} = \overline{\mathcal{E}}' - \overline{u} \times \overline{B}$ into (4.14), we obtain another useful integral form of Faraday's law as

$$\oint_l \overline{\mathcal{E}}' \cdot d\overline{r} = -\int_s \frac{\partial \overline{B}}{\partial t} \cdot d\overline{s} + \oint_l (\overline{u} \times \overline{B}) \cdot d\overline{r}. \tag{4.15}$$

$$\underbrace{\phantom{\oint_l \overline{\mathcal{E}}' \cdot d\overline{r}}}_{\mathcal{V}}$$

Note that (4.12), (4.14), and (4.15) are all equivalent Faraday's laws that are good for a moving loop. Here, $\frac{d}{dt} \int_s \overline{B} \cdot d\overline{s} = \int_s \frac{\partial \overline{B}}{\partial t} \cdot d\overline{s} - \oint_l (\overline{u} \times \overline{B}) \cdot d\overline{r}$ since $\overline{s}$ varies in time. If the loop becomes stationary ($\overline{u} = 0$), then $\frac{d}{dt} \int_s \overline{B} \cdot d\overline{s} = \int_s \frac{\partial \overline{B}}{\partial t} \cdot d\overline{s}$, thus (4.12), (4.14), and (4.15) all reduce to

$$\oint_l \overline{\mathcal{E}} \cdot d\overline{r} = -\frac{d}{dt} \int_s \overline{B} \cdot d\overline{s}. \tag{4.16}$$

Example 4.1 Emf due to a time-varying magnetic flux density.

Figure 4.6 shows a conducting bar sliding over conducting parallel lines on the x-y plane. This bar moves along the x-direction with a constant velocity $\overline{u}$ $(= \hat{x}u)$ while $\overline{B}$ $(= \hat{z}B_0 \cos \omega t)$ passes through the loop. Determine the emf induced on a conducting loop $(a \times b)$ where $b = ut$ $(t > 0)$.

Solution:

1. **Approach 1:** We first choose a path $[(1) \rightarrow (2) \rightarrow (3) \rightarrow (4) \rightarrow (5) \rightarrow (6) \rightarrow (1)]$. The emf induced around the loop is

Fig. 4.6 Emf due to a
time-varying magnetic flux
density $\bar{B}$

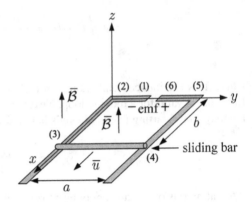

$$V = -\frac{d}{dt} \int_s \bar{B} \cdot d\bar{s} \tag{4.17}$$

where s denotes a rectangular loop area $a \times b$. Letting $d\bar{s} = \hat{z}\, dx\, dy$ and $b = ut$, we get

$$V = -\frac{d}{dt} \left(\int_0^a dy \int_0^{ut} B_0 \cos \omega t\, dx \right)$$

$$= -\frac{d}{dt} (a B_0 ut \cos \omega t)$$

$$= a B_0 (\omega b \sin \omega t - u \cos \omega t). \tag{4.18}$$

If a path $\left[(6) \rightarrow (5) \rightarrow (4) \rightarrow (3) \rightarrow (2) \rightarrow (1) \rightarrow (6) \right]$ were chosen, the emf should be the negative value of (4.18). This second choice should not pose any problem since the ultimate induced current flow remains the same no matter which direction we choose for the path.

2. **Approach 2:** The emf is also given by

$$V = -\underbrace{\int_s \frac{\partial \bar{B}}{\partial t} \cdot d\bar{s}}_{I_1} + \underbrace{\oint_l (\bar{u} \times \bar{B}) \cdot d\bar{r}}_{I_2}. \tag{4.19}$$

Choosing $d\bar{s} = \hat{z}\, dx\, dy$, we evaluate I_1 as

$$I_1 = \omega a b B_0 \sin \omega t. \tag{4.20}$$

Since $\bar{u} = \hat{x} u$ and $d\bar{r} = \hat{y} dy$ for the sliding bar, I_2 gives

$$I_2 = -\int_0^a u B_0 \cos \omega t \, dy = -u B_0 a \cos \omega t. \tag{4.21}$$

Therefore the emf is $V = a B_0(\omega b \sin \omega t - u \cos \omega t)$.

4.2 Inductance

The inductance concept is useful for the analysis of circuits such as current-carrying wire loops. First consider two conducting wire loops in Fig. 4.7. A closed loop l_1 carrying a time-varying current $\mathcal{I}_1$ is placed near another loop l_2 carrying a current $\mathcal{I}_2$, where s_1 and s_2 denote the surfaces enclosed by l_1 and l_2, respectively. A current $\mathcal{I}_1$ in l_1 produces a time-varying magnetic flux density $\bar{B}_1$. According to Faraday's law, $\bar{B}_1$ induces an electromotive force $V = -\dfrac{d\Phi_{12}}{dt}$ in l_2, where the magnetic flux is

$$\Phi_{12} = \int_{s_2} \bar{B}_1 \cdot d\bar{s}_2. \tag{4.22}$$

The current $\mathcal{I}_1$ in l_1 induces the electromotive force in l_2 by means of the magnetic flux density $\bar{B}_1$. What is the relation between the time-varying $\bar{B}_1$ and $\mathcal{I}_1$? A partial answer lies in the Biot-Savart law. The Biot-Savart law states that the static magnetic flux density is directly proportional to the time-invariant steady current. If $\mathcal{I}_1$ varies slowly enough to be considered time-invariant, the Biot-Savart law applies to this slowly time-varying case. Therefore, Φ_{12} is considered to be directly proportional to $\mathcal{I}_1$ as

$$\Phi_{12} = L_{12}\mathcal{I}_1 \tag{4.23}$$

where the proportionality constant L_{12} is referred to as the mutual inductance between l_1 and l_2. When l_2 has N_2 turns, it is convenient to introduce a magnetic flux linkage $\Lambda_{12} = N_2\Phi_{12}$. The mutual inductance is defined as

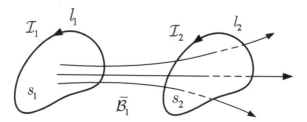

Fig. 4.7 Two current loops l_1 and l_2

$$L_{12} = \frac{\Lambda_{12}}{\mathcal{I}_1}. \tag{4.24}$$

definition of mutual inductance

Neumann's formula for mutual inductance: Consider two neighboring current loops in Fig. 4.8. Since $\mathcal{I}_1$ is slowly varying, it is acceptable to use the magnetic vector potential, which was originally defined for time-invariant steady currents. The magnetic vector potential $\overline{A}$ at $\overline{r}_2$ due to a time-varying current density $\overline{\mathcal{J}}'$ at $\overline{r}_1$ is approximately

$$\overline{A} = \frac{\mu_0}{4\pi} \int_{v'} \frac{\overline{\mathcal{J}}'}{R} \, dv'$$

$$= \frac{\mu_0 \mathcal{I}_1}{4\pi} \oint_{l_1} \frac{d\overline{r}_1}{R} \tag{4.25}$$

where $R = |\overline{r}_2 - \overline{r}_1|$. The magnetic flux density at $\overline{r}_2$ is

$$\overline{B}_1 = \nabla \times \overline{A}. \tag{4.26}$$

The magnetic flux linkage becomes

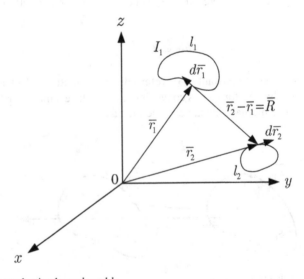

Fig. 4.8 Two conducting loops l_1 and l_2

$$\Lambda_{12} = \int_{s_2} \overline{B}_1 \cdot d\overline{s}_2$$

$$= \int_{s_2} \nabla \times \overline{A} \cdot d\overline{s}_2$$

$$= \oint_{l_2} \overline{A} \cdot d\overline{r}_2. \tag{4.27}$$

We have used Stokes's theorem to obtain (4.27). Substituting (4.25) into (4.27) yields Neumann's formula for the mutual inductance as

$$L_{12} = \frac{\Lambda_{12}}{\mathcal{I}_1}$$

$$= \frac{\mu_0}{4\pi} \oint_{l_2} \oint_{l_1} \frac{d\overline{r}_1 \cdot d\overline{r}_2}{R}. \tag{4.28}$$

Since interchanging the indices 1 and 2 in (4.28) does not alter the value of L_{12}, the mutual inductance is said to be reciprocal ($L_{12} = L_{21}$).

Self-inductance: The current in l_1 also induces the electromotive force by itself and this self-induction phenomenon is explained in terms of the self-inductance. When a conducting wire loop l_1 consists of N_1 turns, the self-inductance L_{11} (or inductance L) of l_1 is defined as

$$L_{11}(= L) = \frac{\Lambda_{11}}{\mathcal{I}_1} \tag{4.29}$$

definition of self-inductance

where

$$\Lambda_{11} = N_1 \int_{s_1} \overline{B}_1 \cdot d\overline{s}_1. \tag{4.30}$$

Example 4.2 Inductance of a solenoid.
Figure 4.9 shows a solenoid with an air core having a circular cross section s ($= \pi a^2$) and a length $l \gg a$. The solenoid consists of tightly wound N-turn loops carrying a current $\mathcal{I}$. Evaluate the inductance.

Solution: A solenoid can be modeled as a stack of circular loops. Since $l \gg a$, the N-turn solenoid is assumed infinitely long. If a solenoid is infinitely long, only an

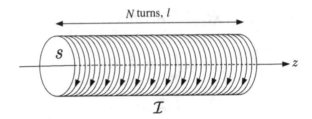

Fig. 4.9 Solenoid with a cross section s and N turns carrying a current $\mathcal{I}$

axial component $\mathcal{B}_z$ of the magnetic flux density $\overline{B}$ exists within the solenoid, while $\overline{B}$ is zero outside the solenoid. We will use Ampère's circuital law to determine $\mathcal{B}_z$ inside the solenoid. In order to apply Ampère's circuital law, we choose a closed rectangular path c $(1 \rightarrow 2 \rightarrow 3 \rightarrow 4 \rightarrow 1)$ encircling the N-turn solenoid, as illustrated in Fig. 4.10. The line integral along the closed path is

$$\oint_c \overline{B} \cdot d\overline{r} =$$

$$\underbrace{\int_1^2 \overline{B} \cdot d\overline{r}}_{\mathcal{B}_z l} + \underbrace{\int_2^3 \overline{B} \cdot d\overline{r}}_{0} + \underbrace{\int_3^4 \overline{B} \cdot d\overline{r}}_{0} + \underbrace{\int_4^1 \overline{B} \cdot d\overline{r}}_{0}. \tag{4.31}$$

Ampère's circuital law gives

$$\oint_c \overline{B} \cdot d\overline{r} = \mu_0 N \mathcal{I} \Longrightarrow \mathcal{B}_z = \frac{\mu_0 N \mathcal{I}}{l}. \tag{4.32}$$

Therefore

$$\Lambda_{11} = N \underbrace{\int_s \overline{B} \cdot d\overline{s}}_{\mathcal{B}_z s} = \frac{\mu_0 N^2 \mathcal{I} s}{l}. \tag{4.33}$$

The inductance is

$$L = \frac{\Lambda_{11}}{\mathcal{I}} = \frac{\mu_0 N^2 s}{l}. \tag{4.34}$$

Example 4.3 Inductance of a coaxial cable.

In Fig. 4.11, determine the inductance of a coaxial cable of length l consisting of two conducting concentric cylinders with radii $\rho = a$ and $\rho = b$ $(a, b \ll l)$. The annular

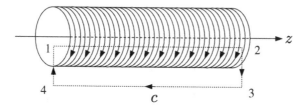

Fig. 4.10 A closed path c surrounding the current

Fig. 4.11 A coaxial cable
with radii $\rho = a$ and $\rho = b$
carrying a current $\mathcal{I}$

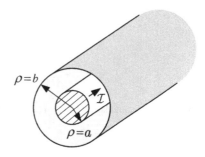

space is filled with a material of permeability μ_0. Assume that the current $\mathcal{I}$ flows
on the conducting cylindrical surfaces at $\rho = a$ and $\rho = b$.

Solution: We first set up the cylindrical coordinates (ρ, ϕ, z) coincident with a coaxial
cable geometry. The magnetic flux density can be obtained from Ampère's circuital
law. The magnetic flux density within a coaxial cable has a ϕ-component as

$$
\mathcal{B}_\phi = \begin{cases} 0 & \text{for } 0 < \rho < a \\[2mm] \dfrac{\mu_0 \mathcal{I}}{2\pi\rho} & \text{for } a < \rho < b. \end{cases} \tag{4.35}
$$

We next consider an annular area $(a < \rho < b)$ through which the magnetic flux density $\mathcal{B}_\phi$ passes. Figure 4.12 illustrates a cross section of rectangular area surrounded
by a current loop. The magnetic flux linkage is

$$
\Lambda_{11} = \int_a^b \frac{\mu_0 \mathcal{I}}{2\pi\rho} l \, d\rho = \frac{\mu_0 \mathcal{I} l}{2\pi} \ln\left(\frac{b}{a}\right). \tag{4.36}
$$

The inductance is

$$
L = \frac{\Lambda_{11}}{\mathcal{I}} = \frac{\mu_0 l}{2\pi} \ln\left(\frac{b}{a}\right). \tag{4.37}
$$

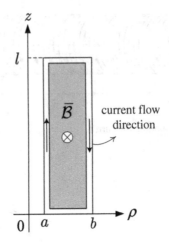

Fig. 4.12 A rectangular area $(b - a) \times l$ where $l \gg a, b$

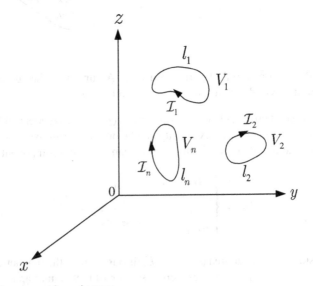

Fig. 4.13 Many current loops in space

4.3 Magnetic Energy

It has been known that energy can be stored in magnetic fields. To derive magnetic energy stored in magnetic fields, we consider a set of current loops l_n ($n = 1, 2, \cdots, N$) placed in a linear medium with $\mu = \mu_0$, as shown in Fig. 4.13. Initially there were no currents and currents are increased from zero to $\mathcal{I}_n$ in l_n. A change in currents inevitably produces magnetic flux density change and electromotive forces V_n in l_n. The electromotive forces oppose current changes from zero to

$\mathcal{I}_n$. To overcome these induced electromotive forces and to maintain $\mathcal{I}_n$, work must be done by supplying a voltage $V_n (= -\mathcal{V}_n)$ to l_n. The work done per unit time is

$$\frac{dW_m}{dt} = \sum_{n=1}^{N} V_n \mathcal{I}_n$$

$$= \sum_{n=1}^{N} \frac{d\Phi_n}{dt} \mathcal{I}_n \tag{4.38}$$

where Φ_n is the magnetic flux that passes through l_n due to currents on l_1 through l_N. We introduce inductances L_{jn} as

$$L_{jn} = \frac{\phi_{jn}}{\mathcal{I}_j} \tag{4.39}$$

where ϕ_{jn} is the magnetic flux that passes through l_n due to a current $\mathcal{I}_j$ on l_j. Note

$$L_{jn} = \begin{cases} \text{mutual inductances} & \text{for } n \neq j \\ \\ \text{self-inductances} & \text{for } n = j. \end{cases} \tag{4.40}$$

Hence the total flux that passes through l_n is

$$\Phi_n = \sum_{j=1}^{N} \phi_{jn} = \sum_{j=1}^{N} L_{jn} \mathcal{I}_j. \tag{4.41}$$

Therefore

$$\frac{dW_m}{dt} = \sum_{n=1}^{N} \sum_{j=1}^{N} L_{jn} \frac{d\mathcal{I}_j}{dt} \mathcal{I}_n. \tag{4.42}$$

We wish to find a work to maintain the final steady-state current $\mathcal{I}_n$. Integrating $\frac{dW_m}{dt}$ with respect to time from 0 to the final state gives

$$W_m = \frac{1}{2} \sum_{n=1}^{N} \underbrace{\sum_{j=1}^{N} L_{jn} \mathcal{I}_j}_{\Phi_n} \mathcal{I}_n. \tag{4.43}$$

This is the magnetic energy that is stored by a set of loops carrying the currents $\mathcal{I}_n$ with the inductances L_{jn}. If the loop is single, W_m can be further simplified. The magnetic energy, stored in an inductor with a self-inductance L, is

$$W_m = \frac{1}{2}L\mathcal{I}^2. \tag{4.44}$$

magnetic energy in terms of self-inductance

Magnetic energy in terms of fields: Next we will derive magnetic energy in terms of magnetic fields. If $\overline{B}_{jn}$ denotes the magnetic flux density through s_n due to $\mathcal{I}_j$, the total flux passing through s_n due to $\mathcal{I}_j$ ($j = 1, 2, \cdots, N$) is

$$
\begin{aligned}
\Phi_n &= \sum_{j=1}^{N} \int_{s_n} \overline{B}_{jn} \cdot d\overline{s} \\
&= \sum_{j=1}^{N} \int_{s_n} (\nabla \times \overline{A}_{jn}) \cdot d\overline{s} \\
&= \sum_{j=1}^{N} \oint_{l_n} \overline{A}_{jn} \cdot d\overline{r}
\end{aligned}
\tag{4.45}
$$

where s_n is any surface enclosed by l_n and $\overline{A}_{jn}$ is the magnetic vector potential due to $\mathcal{I}_j$. Note that Stokes's theorem was used in (4.45). Hence (4.43) is

$$W_m = \frac{1}{2} \sum_{n=1}^{N} \left(\sum_{j=1}^{N} \oint_{l_n} \overline{A}_{jn} \cdot d\overline{r} \right) \mathcal{I}_n. \tag{4.46}$$

So far we have discussed cases of loop currents. To generalize about currents, we will replace $\mathcal{I}_n$ with a volume current density $\overline{\mathcal{J}}$ that occupies a volume v. Using the equivalence $\mathcal{I}_n \, d\overline{r} = \overline{\mathcal{J}} \, dv$, we obtain

$$W_m = \frac{1}{2} \int_v \overline{A} \cdot \overline{\mathcal{J}} \, dv \tag{4.47}$$

where $\overline{A}$ is the magnetic vector potential due to $\overline{\mathcal{J}}$ occupying v. It is possible to extend v to an unbounded space by including an additional space devoid of currents ($\overline{\mathcal{J}} = 0$). Using Ampère's circuital law $\nabla \times \overline{\mathcal{H}} = \overline{\mathcal{J}}$ gives

$$W_m = \frac{1}{2} \int_v \overline{A} \cdot (\nabla \times \overline{\mathcal{H}}) \, dv. \tag{4.48}$$

Further simplification is possible by using the identity

$$\nabla \cdot (\overline{A} \times \overline{\mathcal{H}}) = -\overline{A} \cdot (\nabla \times \overline{\mathcal{H}}) + \overline{\mathcal{H}} \cdot \underbrace{(\nabla \times \overline{A})}_{\overline{B}} . \tag{4.49}$$

Expression (4.48) becomes

$$W_m = \frac{1}{2} \int_v \overline{\mathcal{H}} \cdot \overline{B} \, dv - \frac{1}{2} \underbrace{\int_v \nabla \cdot (\overline{A} \times \overline{\mathcal{H}}) \, dv}_{\oint_s (\overline{A} \times \overline{\mathcal{H}}) \cdot d\overline{s}} . \tag{4.50}$$

Here the divergence theorem has been utilized in (4.50) where s denotes a spherical surface with radius $r \to \infty$. The fields on a spherical surface s decay as $\overline{A} \to 1/r$ and $\overline{\mathcal{H}} \to 1/r^2$ when $r \to \infty$ (see the Biot-Savart law and the definition of magnetic vector potential in Sect. 3.2). Since $d\overline{s}$ varies as r^2,

$$\oint_s (\overline{A} \times \overline{\mathcal{H}}) \cdot d\overline{s} \to 0. \tag{4.51}$$

The magnetic energy stored in magnetic fields becomes

$$W_m = \frac{1}{2} \int \overline{\mathcal{H}} \cdot \overline{B} \, dv \tag{4.52}$$

magnetic energy in terms of fields

where the integration extends over the infinite space.

Example 4.4 Magnetic energy stored in a solenoid.
Evaluate the magnetic energy stored in a long solenoid in Fig. 4.9.

Solution: The magnetic flux density in a solenoid is $B_z = \mu_0 N \mathcal{I}/l$. The stored energy is

$$W_m = \frac{1}{2} \int \frac{B_z^2}{\mu_0} \, dv$$

$$= \frac{1}{2} \underbrace{\frac{\mu_0 N^2 s}{l}}_{L} \mathcal{I}^2$$

$$= \frac{1}{2} L \mathcal{I}^2 \tag{4.53}$$

where L is the self-inductance of a solenoid.

Fig. 4.14 A rotating loop in a static magnetic flux density $\overline{B}$

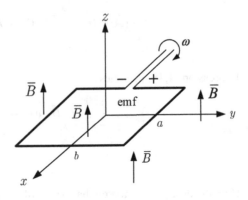

4.4 Problems for Chapter 4

1. Starting from $\nabla \cdot \overline{B} = 0$, show that the contour shape of $\overline{s}$ in (4.3) may be arbitrary. **Hint**: Apply the divergence theorem to $\nabla \cdot \overline{B} = 0$.

2. Show that the voltage between two points ($v_{12} = -\int_2^1 \overline{\mathcal{E}} \cdot d\overline{r}$) is not unique but path-dependent in general for time-varying fields.

3. Figure 4.14 shows a rotating rectangular loop of size ($a \times b$) placed in a static magnetic flux density $\overline{B}$ ($= \hat{z}B$). At time $t = 0$, the surface of the loop is on

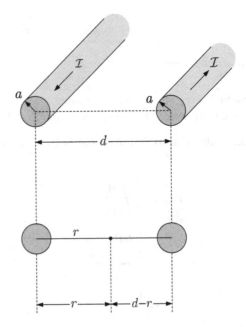

Fig. 4.15 Parallel wires carrying currents $\mathcal{I}$

the x-y plane. Find the emf induced along the loop that rotates with a constant angular velocity ω.

Hint: Use Faraday's law.

4. Figure 4.15 shows parallel wires of infinite length and radius a separated by a spacing d ($d \gg a$) in air. The parallel wires carry equal but opposite currents I. Determine the self-inductance per unit length of the parallel wires.

Hint: Use Ampère's circuital law to approximately determine the magnetic flux density at r.

5. Figure 4.16 depicts two coaxial circular wires of radii a and b separated by a distance h ($h \gg a$ and $h \gg b$). Find the mutual inductance between two wires.

Hint: Since $h \gg a$ and $h \gg b$, the magnetic flux density through the upper circular loop is approximately constant and points in the z-direction.

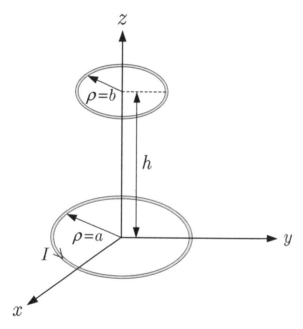

Fig. 4.16 Two coaxial circular wires of radii a and b

Chapter 5
Maxwell's Equations

5.1 Continuity Equation and Ampère's Law

5.1.1 Continuity Equation

Time-varying currents and charges are related by the continuity equation in electro-dynamics (Fig. 5.1). We will derive the continuity equation by using the principle of charge conservation. Figure 5.2 shows a charge Q that occupies a volume v surrounded by a closed surface s. When a differential charge dQ leaks out through the surface s over a differential time period dt, the leaking current $\mathcal{I}$ is given by

$$\mathcal{I} = -\frac{dQ}{dt} \ . \tag{5.1}$$

The negative sign implies that a creation of current comes from an annihilation of charge. The current and the charge are given by

$$\mathcal{I} = \oint_s \overline{\mathcal{J}} \cdot d\overline{s} \tag{5.2}$$

$$Q = \int_v \rho_v \, dv \tag{5.3}$$

where $\overline{\mathcal{J}}$ and ρ_v are the current density and the volume charge density, respectively. Substituting (5.2) and (5.3) into (5.1) and applying the divergence theorem $\oint_s \overline{\mathcal{J}} \cdot d\overline{s} = \int_v \nabla \cdot \overline{\mathcal{J}} \, dv$, we obtain

H. J. Eom, *Primary Theory of Electromagnetics*, Power Systems,
DOI: 10.1007/978-94-007-7143-7_5, © Springer Science+Business Media Dordrecht 2013

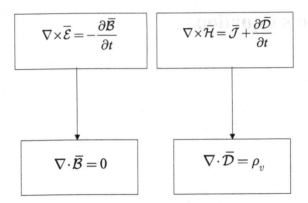

$$\nabla \times \bar{\mathcal{E}} = -\frac{\partial \bar{B}}{\partial t} \qquad \nabla \times \bar{\mathcal{H}} = \bar{J} + \frac{\partial \bar{D}}{\partial t}$$

$$\nabla \cdot \bar{B} = 0 \qquad \nabla \cdot \bar{D} = \rho_v$$

Fig. 5.1 Fundamental equations for time-varying fields

Fig. 5.2 A charge $\mathcal{Q}$
surrounded by a surface s

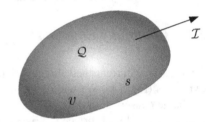

$$\nabla \cdot \bar{J} = -\frac{\partial \rho_v}{\partial t}. \tag{5.4}$$

continuity equation

The continuity equation states that a decrease in charge density $-\partial \rho_v / \partial t$ comes at the expense of an increase in current density $\nabla \cdot \bar{J}$; the continuity equation amounts to a charge conservation law.

Example 5.1 Charges in conductors.
Show that charges within conductors vanish very quickly and are considered zero for all practical purposes.

Solution: The current density $\bar{J}$ flowing in a conductor of conductivity σ obeys Ohm's law

$$\bar{J} = \sigma \bar{\mathcal{E}}. \tag{5.5}$$

Substituting Ohm's law into the continuity equation yields

$$\sigma \nabla \cdot \overline{\mathcal{E}} = -\frac{\partial \rho_v}{\partial t}.$$ (5.6)

Since Gauss's law is given by $\nabla \cdot \overline{\mathcal{E}} = \frac{\rho_v}{\epsilon}$, (5.6) becomes

$$\frac{\partial \rho_v}{\partial t} + \frac{\sigma}{\epsilon} \rho_v = 0$$ (5.7)

where the solution is

$$\rho_v = \rho_0 \exp\left(-\frac{\sigma}{\epsilon} t\right).$$ (5.8)

For copper with $\sigma = 5.8 \times 10^7$ (S/m) and $\epsilon = 8.85 \times 10^{-12}$ (F/m), the time constant is extremely small $\epsilon/\sigma = 1.53 \times 10^{-19}$ (s); charges within conductors disappear instantaneously. Therefore charges cannot exist inside good conductors.

5.1.2 Ampère's Law

Ampère's circuital law $\nabla \times \overline{H} = \overline{J}$ is valid when $\overline{H}$ and $\overline{J}$ are both time-invariant. Is Ampère's circuital law still valid for time-varying $\overline{\mathcal{H}}$ and $\overline{\mathcal{J}}$? The answer comes from the divergence of Ampère's circuital law:

$$\underbrace{\nabla \cdot (\nabla \times \overline{\mathcal{H}})}_{0} = \underbrace{\nabla \cdot \overline{\mathcal{J}}}_{-\frac{\partial \rho_v}{\partial t}}.$$ (5.9)

Apparently this contradicts the continuity equation in time-varying cases: the relation $\nabla \times \overline{\mathcal{H}} = \overline{\mathcal{J}}$ should be approximately valid only when fields are slowly time-varying $\partial/\partial t \approx 0$. To correct this contradiction, James Clerk Maxwell cleverly added a term $\frac{\partial \overline{D}}{\partial t}$ to Ampère's circuital law to obtain

$$\nabla \times \overline{\mathcal{H}} = \overline{\mathcal{J}} + \frac{\partial \overline{D}}{\partial t}.$$ (5.10)

Ampère's law in differential form

The term $\frac{\partial \overline{D}}{\partial t}$ is referred to as a displacement current density. The validity of Ampère's law was confirmed experimentally and is considered one of fundamental laws in electrodynamics. Ampère's law relates the electromagnetic fields to the

Fig. 5.3 Circuit elements for
various currents

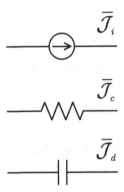

current density. The current density $\overline{\mathcal{J}}$ can be decomposed into two different components, $\overline{\mathcal{J}}_i$ and $\overline{\mathcal{J}}_c$, as

$$\overline{\mathcal{J}} = \overline{\mathcal{J}}_i + \overline{\mathcal{J}}_c \tag{5.11}$$

where $\overline{\mathcal{J}}_i$ is an impressed current density and $\overline{\mathcal{J}}_c$ is a conduction current density given by $\sigma\overline{\mathcal{E}}$. The circuit elements in Fig. 5.3 give simplified views of different current components. The current density supplied by a source is the impressed current density $\overline{\mathcal{J}}_i$, the current density flowing through a conductor is the conduction current density $\overline{\mathcal{J}}_c$, and the current density flowing through a capacitor is called the displacement current density $\overline{\mathcal{J}}_d = \dfrac{\partial \overline{\mathcal{D}}}{\partial t}$.

5.2 Maxwell's Equations

5.2.1 Time-Varying Forms

Time-varying electromagnetic fields are known to obey Faraday's law and Ampère's law. A collection of Faraday's law, Ampère's law, and two Gauss's laws is called Maxwell's equations. Gauss's laws can be analytically derived from Faraday's law and Ampère's law. For instance, applying the divergence to Ampère's law yields

$$\underbrace{\nabla \cdot \left(\nabla \times \overline{\mathcal{H}} \right)}_{0} = \underbrace{\nabla \cdot \overline{\mathcal{J}}}_{-\frac{\partial \rho_v}{\partial t}} + \frac{\partial}{\partial t} \nabla \cdot \overline{\mathcal{D}} \tag{5.12}$$

where $\nabla \cdot \left(\nabla \times \overline{\mathcal{H}} \right)$ should be zero for any arbitrary vector $\overline{\mathcal{H}}$, and the continuity equation has been utilized. Integrating (5.12) with respect to t and setting the integration constant to zero, we obtain Gauss's law for time-varying fields:

$$\nabla \cdot \overline{D} = \rho_v .\tag{5.13}$$

Here the integration constant was set to zero since the field was supposed to vanish in its past history. Similarly, applying the divergence to Faraday's law, we obtain

$$\underbrace{\nabla \cdot \left(\nabla \times \overline{\mathcal{E}} \right)}_{0} = -\frac{\partial}{\partial t} \nabla \cdot \overline{B}.\tag{5.14}$$

Hence the magnetic Gauss's law for time-varying fields is

$$\nabla \cdot \overline{B} = 0.\tag{5.15}$$

Maxwell's equations are

$$\nabla \times \overline{\mathcal{E}} = -\frac{\partial \overline{B}}{\partial t} \quad (Faraday's\ law)\tag{5.16}$$

$$\nabla \times \overline{\mathcal{H}} = \overline{\mathcal{J}} + \frac{\partial \overline{D}}{\partial t} \quad (Ampère's\ law)\tag{5.17}$$

$$\nabla \cdot \overline{D} = \rho_v \quad (Gauss's\ law)\tag{5.18}$$

$$\nabla \cdot \overline{B} = 0 \quad (magnetic\ Gauss's\ law)\tag{5.19}$$

Maxwell's equations in differential form

where $\overline{\mathcal{J}}$ and ρ_v are the volume current density of free charges and the volume charge density of free charges, respectively. For a medium with a permittivity ϵ and a permeability μ, the constitutive relations are given by

$$\overline{D} = \epsilon \overline{\mathcal{E}}\tag{5.20}$$

$$\overline{B} = \mu \overline{\mathcal{H}}.\tag{5.21}$$

Maxwell's equations govern the electric and magnetic fields, $\overline{\mathcal{E}}$ and $\overline{\mathcal{H}}$. In particular, two independent equations, Faraday's law and Ampère's law, are usually solved to determine $\overline{\mathcal{E}}$ and $\overline{\mathcal{H}}$. We can rewrite Maxwell's equations in integral form. This can be achieved by applying Stokes's theorem to Faraday's law and Ampère's law, and by applying the divergence theorem to Gauss's laws. The results are summarized as

$$\oint_l \overline{\mathcal{E}} \cdot d\overline{r} = -\int_s \frac{\partial \overline{B}}{\partial t} \cdot d\overline{s} \quad (Faraday's\ law) \tag{5.22}$$

$$\oint_l \overline{\mathcal{H}} \cdot d\overline{r} = \int_s \overline{\mathcal{J}} \cdot d\overline{s} + \int_s \frac{\partial \overline{D}}{\partial t} \cdot d\overline{s} \quad (Ampère's\ law) \tag{5.23}$$

$$\oint_s \overline{D} \cdot d\overline{s} = \int_v \rho_v \, dv \quad (Gauss's\ law) \tag{5.24}$$

$$\oint_s \overline{B} \cdot d\overline{s} = 0 \quad (magnetic\ Gauss's\ law). \tag{5.25}$$

Maxwell's equations in integral form

The notations l, s, and v follow usual conventions associated with Stokes's and divergence theorems. It should be noted that time-varying electromagnetic fields are waves that propagate in space to deliver electromagnetic energy. Wireless and optical communications utilize electromagnetic waves to transmit information from one location to another. Propagating electromagnetic waves will be discussed in greater detail in Chap. 6.

Lorentz force equation: It is known that charges moving in electromagnetic field experience forces. If a charge q moves at an instantaneous velocity $\overline{u}$ in the fields of $\overline{\mathcal{E}}$ and $\overline{B}$, the charge experiences a Lorentz force $\overline{\mathcal{F}}$ as follows:

$$\overline{\mathcal{F}} = q(\overline{\mathcal{E}} + \overline{u} \times \overline{B}). \tag{5.26}$$

Lorentz force equation

Lorentz force equation is a fundamental law that supplements Maxwell's equations. Note that the validity of Lorentz force equation has been well established.

5.2.2 Boundary Conditions

Time-varying fields between two different media follow a certain rule, which is known as boundary conditions. We consider electromagnetic fields near the boundary, as shown in Fig. 5.4. The boundary conditions for time-varying fields are the same as the boundary conditions for static fields that were derived under the static assumption. Since the procedure for time-varying fields is essentially identical with that for static fields, we will simply show the final results. Applying Faraday's law and Ampère's law to the boundary fields, we write the boundary conditions for time-varying fields as

Fig. 5.4 A boundary between
two media

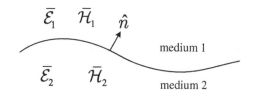

$$\hat{n} \times \left(\overline{\mathcal{E}}_1 - \overline{\mathcal{E}}_2 \right) = 0 \tag{5.27}$$

$$\hat{n} \times \left(\overline{\mathcal{H}}_1 - \overline{\mathcal{H}}_2 \right) = \overline{\mathcal{J}}_s \tag{5.28}$$

where $\overline{\mathcal{J}}_s$ is the surface current density impressed at the boundary. Here $\hat{n}$ is a unit
vector normal to the boundary. Equation (5.27) states that the tangential components
of $\overline{\mathcal{E}}$ are continuous across the boundary. Equation (5.28) states that the tangential
components of $\overline{\mathcal{H}}$ are discontinuous across the boundary by the amount $\overline{\mathcal{J}}_s$. Auxiliary
boundary conditions are also obtained from the two Gauss's laws as

$$\hat{n} \cdot \left(\overline{\mathcal{D}}_1 - \overline{\mathcal{D}}_2 \right) = \rho_s \tag{5.29}$$

$$\hat{n} \cdot \left(\overline{\mathcal{B}}_1 - \overline{\mathcal{B}}_2 \right) = 0 \tag{5.30}$$

where ρ_s is the surface charge density present at the boundary. We note that $\hat{n} \cdot \overline{\mathcal{D}}_1$
refers to the normal component of electric flux density at the boundary in medium 1.
Equation (5.29) states that the normal components of $\overline{\mathcal{D}}$ are discontinuous across the
boundary by the amount ρ_s. Equation (5.30) states that the normal components of $\overline{\mathcal{B}}$
are continuous across the boundary.

Boundary conditions for a perfect electric conductor: When medium 2 is perfectly conducting (perfect electric conductor, PEC: conductivity $\sigma \rightarrow \infty$), the above
boundary conditions should be reconsidered. Electric fields inside the PEC are zero
$(\overline{\mathcal{E}}_2 = \overline{\mathcal{D}}_2 = 0)$ and charges stay at the PEC surface. If $\overline{\mathcal{E}}_2$ were nonzero, the current density $(\overline{\mathcal{J}} = \sigma \overline{\mathcal{E}}_2)$ would become infinite! From Faraday's law we conclude
$\overline{\mathcal{B}}_2 = \overline{\mathcal{H}}_2 = 0$ inside the PEC as long as fields are time-varying. Hence, all time-
varying fields within PEC must be zero. The boundary conditions between a dielectric
(medium 1) and a PEC (medium 2) are written as

$$\hat{n} \times \overline{\mathcal{E}}_1 = 0 \tag{5.31}$$

$$\hat{n} \times \overline{\mathcal{H}}_1 = \overline{\mathcal{J}}_s \tag{5.32}$$

$$\hat{n} \cdot \overline{\mathcal{D}}_1 = \rho_s \tag{5.33}$$

$$\hat{n} \cdot \overline{B}_1 = 0 \tag{5.34}$$

where $\overline{\mathcal{J}}_s$ is the surface current density on the PEC surface, and ρ_s is the surface charge density on the PEC surface.

5.2.3 Static Limit

Time-varying fields become static when $\partial/\partial t \rightarrow 0$. Maxwell's equations in static limit yield the electrostatic and magnetostatic results, which were derived from Coulomb's law in Chap. 2 and from Ampère's law of force in Chap. 3. We will consider two static cases separately as follows.

1. For electrostatic fields

$$\nabla \times \overline{E} = 0 \tag{5.35}$$

$$\nabla \cdot \overline{D} = \rho_v \tag{5.36}$$

where $\overline{E}, \overline{D}$, and ρ_v are all time-invariant quantities.
2. For magnetostatic fields

$$\nabla \times \overline{H} = \overline{J} \tag{5.37}$$

$$\nabla \cdot \overline{B} = 0 \tag{5.38}$$

where $\overline{H}, \overline{B}$, and $\overline{J}$ are all time-invariant quantities.

The continuity equation becomes

$$\nabla \cdot \overline{J} = \underbrace{-\frac{\partial \rho_v}{\partial t}}_{0}. \tag{5.39}$$

In static limit, a current density $\overline{J}$ and a charge density ρ_v behave as independent sources: $\overline{J}$ produces $\overline{H}$ and ρ_v produces $\overline{E}$ independently. There is no interdependence between $\overline{E}$ and $\overline{H}$.

5.3 Poynting's Theorem

Electromagnetic fields can deliver power. An electromagnetic power delivery is explained in terms of Poynting's theorem. Derivation of Poynting's theorem begins with Faraday's law and Ampère's law

$$\nabla \times \overline{\mathcal{E}} = -\mu \frac{\partial \overline{\mathcal{H}}}{\partial t} \tag{5.40}$$

$$\nabla \times \overline{\mathcal{H}} = \sigma \overline{\mathcal{E}} + \epsilon \frac{\partial \overline{\mathcal{E}}}{\partial t} + \overline{\mathcal{J}}_i. \tag{5.41}$$

Taking dot products of (5.40) and (5.41) with $\overline{\mathcal{H}}$ and $\overline{\mathcal{E}}$, respectively, and subtracting each other, we obtain

$$\overbrace{\overline{\mathcal{E}} \cdot (\nabla \times \overline{\mathcal{H}}) - \overline{\mathcal{H}} \cdot (\nabla \times \overline{\mathcal{E}})}^{-\nabla \cdot (\overline{\mathcal{E}} \times \overline{\mathcal{H}})} =$$

$$\overline{\mathcal{E}} \cdot \overline{\mathcal{J}}_i + \sigma \underbrace{\overline{\mathcal{E}} \cdot \overline{\mathcal{E}}}_{|\overline{\mathcal{E}}|^2} + \epsilon \underbrace{\overline{\mathcal{E}} \cdot \frac{\partial \overline{\mathcal{E}}}{\partial t}}_{\frac{1}{2} \frac{\partial |\overline{\mathcal{E}}|^2}{\partial t}} + \mu \underbrace{\overline{\mathcal{H}} \cdot \frac{\partial \overline{\mathcal{H}}}{\partial t}}_{\frac{1}{2} \frac{\partial |\overline{\mathcal{H}}|^2}{\partial t}}. \tag{5.42}$$

This is rewritten as

$$-\nabla \cdot (\overline{\mathcal{E}} \times \overline{\mathcal{H}}) =$$

$$\overline{\mathcal{E}} \cdot \overline{\mathcal{J}}_i + \sigma |\overline{\mathcal{E}}|^2 + \frac{\epsilon}{2} \frac{\partial |\overline{\mathcal{E}}|^2}{\partial t} + \frac{\mu}{2} \frac{\partial |\overline{\mathcal{H}}|^2}{\partial t}. \tag{5.43}$$

We next consider a volume v surrounded by a surface s, as shown in Fig. 5.5. Applying a volume integral to (5.43) and utilizing the divergence theorem, we obtain

$$-\underbrace{\int_v \overline{\mathcal{E}} \cdot \overline{\mathcal{J}}_i dv}_{\mathcal{P}_i} = \underbrace{\oint_s (\overline{\mathcal{E}} \times \overline{\mathcal{H}}) \cdot d\overline{s}}_{\mathcal{P}_c} + \underbrace{\int_v \sigma |\overline{\mathcal{E}}|^2 dv}_{\mathcal{P}_d}$$

$$+ \frac{\partial}{\partial t} \underbrace{\int_v \frac{\epsilon}{2} |\overline{\mathcal{E}}|^2 dv}_{\mathcal{W}_e} + \frac{\partial}{\partial t} \underbrace{\int_v \frac{\mu}{2} |\overline{\mathcal{H}}|^2 dv}_{\mathcal{W}_m}. \tag{5.44}$$

$$Poynting's\ theorem$$

Table 5.1 illustrates the physical interpretations of the terms in Poynting's theorem. In terms of a power conservation law, $\mathcal{P}_c$ is regarded as a power flow leaving the surface s surrounding the volume v, as shown in Fig. 5.5. The term $\overline{\mathcal{E}} \times \overline{\mathcal{H}}$, called a Poynting vector, represents a power density carried by the fields, $\overline{\mathcal{E}}$ and $\overline{\mathcal{H}}$.

Fig. 5.5 Power flow $\mathcal{P}_c$
through a closed surface s

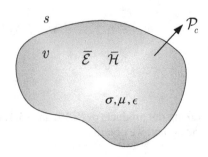

Table 5.1 Power and energy relations

Symbol	Description
$\mathcal{P}_i$	Power supplied by the source $\overline{\mathcal{J}}_i$ within a volume v
$\mathcal{P}_d$	Power dissipation due to the medium conductivity σ
$\mathcal{W}_e$	Stored electric energy
$\mathcal{W}_m$	Stored magnetic energy

5.4 Phasors for Time-Harmonic Fields

5.4.1 Maxwell's Equations in Phasor Form

In most electromagnetic applications, time-varying fields take sinusoidal forms, usually known as time-harmonic. Time-harmonic fields are real functions of position $\overline{r}$
and time t. We write a time-harmonic (time-varying) field with an angular frequency
ω ($= 2\pi f$, f: frequency) as

$$\overline{\mathcal{E}} = \hat{x} E_1 \cos{(\omega t + \phi_1)}$$
$$+ \hat{y} E_2 \cos{(\omega t + \phi_2)}$$
$$+ \hat{z} E_3 \cos{(\omega t + \phi_3)} \tag{5.45}$$

where (E_1, E_2, E_3) and (ϕ_1, ϕ_2, ϕ_3) are the amplitudes and phases, which are
functions of position $\overline{r}$. It is possible to rewrite (5.45) as

$$\overline{\mathcal{E}} = \mathrm{Re}\left[\underbrace{\left(\hat{x} E_1 \, e^{j\phi_1} + \hat{y} E_2 \, e^{j\phi_2} + \hat{z} E_3 \, e^{j\phi_3} \right)}_{\overline{E}(\overline{r})} e^{j\omega t} \right] \tag{5.46}$$

where the symbol Re $(\cdot)$ denotes the real part of $(\cdot)$, and the symbol j denotes the
imaginary unit satisfying the relation $j^2 = -1$. Here, $\overline{E}(\overline{r})$ is called a vector phasor,
or simply a phasor. The relation

$$\bar{\mathcal{E}} = \text{Re}\left[\overline{E}(\bar{r})\, e^{j\omega t}\right] \tag{5.47}$$

enables us to recover a time-varying form from the corresponding phasor. The differentiation of $\bar{\mathcal{E}}$ with respect to time is

$$\frac{\partial \bar{\mathcal{E}}}{\partial t} = \text{Re}\left[j\omega\overline{E}(\bar{r})\, e^{j\omega t}\right]. \tag{5.48}$$

The phasor representation for $\dfrac{\partial \bar{\mathcal{E}}}{\partial t}$ is shown to be $j\omega\overline{E}(\bar{r})$, indicating that the replacement $\dfrac{\partial}{\partial t} \to j\omega$ is possible. In the remainder of this text, the notation $\overline{E}$ is used to represent a phasor $\overline{E}(\bar{r})$ in time-harmonic cases unless specified otherwise. Maxwell's equations in terms of phasors $(\overline{E},\ \overline{B},\ \overline{H},\ \overline{D},\ \overline{J},$ and $\rho_v)$ are rewritten as

$$\nabla \times \overline{E} = -j\omega\overline{B} \quad (Faraday's\ law) \tag{5.49}$$

$$\nabla \times \overline{H} = \overline{J} + j\omega\overline{D} \quad (Ampère's\ law) \tag{5.50}$$

$$\nabla \cdot \overline{D} = \rho_v \quad (Gauss's\ law) \tag{5.51}$$

$$\nabla \cdot \overline{B} = 0 \quad (magnetic\ Gauss's\ law). \tag{5.52}$$

Maxwell's equations in phasor form

5.4.2 Poynting's Theorem in Phasor Form

It would be instructive to revisit Poynting's theorem when fields are time-harmonic. Maxwell's curl equations for time-harmonic fields are written in terms of phasors as

$$\nabla \times \overline{E} = -j\omega\mu\overline{H} \tag{5.53}$$

$$\nabla \times \overline{H}^* = (\sigma - j\omega\epsilon)\overline{E}^* + \overline{J}_i^* \tag{5.54}$$

where the symbol $(\cdot)^*$ denotes a complex conjugate of $(\cdot)$. In what follows, the reason for taking the complex conjugate will be apparent. We then have

$$\overline{E} \cdot (\nabla \times \overline{H}^*) - \overline{H}^* \cdot (\nabla \times \overline{E})$$

$$= \overline{E} \cdot \overline{J}_i^* + \sigma \overline{E} \cdot \overline{E}^* - j\omega\epsilon\overline{E} \cdot \overline{E}^* + j\omega\mu\overline{H} \cdot \overline{H}^*. \tag{5.55}$$

Using the vector relation $\nabla \cdot \left(\overline{E} \times \overline{H}^*\right) = \overline{H}^* \cdot \left(\nabla \times \overline{E}\right) - \overline{E} \cdot \left(\nabla \times \overline{H}^*\right)$, we rewrite (5.55) as

$$-\nabla \cdot \left(\frac{1}{2}\overline{E} \times \overline{H}^*\right) = \frac{1}{2}\overline{E} \cdot \overline{J}_i^* + \frac{1}{2}\sigma\overline{E} \cdot \overline{E}^*$$

$$-2j\omega\left(\frac{1}{4}\epsilon\overline{E} \cdot \overline{E}^* - \frac{1}{4}\mu\overline{H} \cdot \overline{H}^*\right). \tag{5.56}$$

We apply a volume integral to (5.56) to obtain

$$-\underbrace{\frac{1}{2}\int_v \overline{E} \cdot \overline{J}_i^* \, dv}_{P_i} = \underbrace{\oint_s \frac{1}{2}\left(\overline{E} \times \overline{H}^*\right) \cdot d\overline{s}}_{P_c} + \underbrace{\frac{1}{2}\int_v \sigma\overline{E} \cdot \overline{E}^* \, dv}_{P_d}$$

$$- 2j\omega\underbrace{\int_v \frac{1}{4}\epsilon\overline{E} \cdot \overline{E}^* \, dv}_{W_e} + 2j\omega\underbrace{\int_v \frac{1}{4}\mu\overline{H} \cdot \overline{H}^* \, dv}_{W_m} \tag{5.57}$$

which is Poynting's theorem for time-harmonic fields:

$$P_i = P_c + P_d - j2\omega(W_e - W_m). \tag{5.58}$$

Note that P_i is the complex power supplied by the sources $\overline{J}_i^*$, P_d is the power dissipated in heat due to a medium conductivity σ, and W_e and W_m are the time-average electric and magnetic energy, respectively. Here P_c denotes the complex power that flows through the surface s. When electromagnetic fields are time-harmonic, it is more meaningful to use a time-average power flow. The time-average power density, carried by time-harmonic electromagnetic fields, is obtained by averaging the instantaneous power density $\overline{\mathcal{E}} \times \overline{\mathcal{H}}$ over a time period T. For instance, if

$$\overline{\mathcal{E}} = \hat{x}E_x \cos(\omega t + \phi_1) \tag{5.59}$$

$$\overline{\mathcal{H}} = \hat{y}H_y \cos(\omega t + \phi_2) \tag{5.60}$$

then

$$\overline{S}_{av} = \frac{1}{T}\int_0^T \overline{\mathcal{E}} \times \overline{\mathcal{H}} \, dt$$

$$= \frac{\hat{z}E_x H_y}{T}\int_0^T \cos(\omega t + \phi_1) \cos(\omega t + \phi_2) \, dt$$

$$= \underbrace{\frac{\hat{z} E_x H_y}{2} \cos{(\phi_1 - \phi_2)}}_{\frac{1}{2} \text{Re} \left(\overline{E} \times \overline{H}^*\right)} \tag{5.61}$$

where $\overline{E}$ and $\overline{H}$ are the phasor representations for the time-varying fields $\mathcal{E}$ and $\mathcal{H}$, respectively. The symbol Re $(\cdot)$ denotes the real part of $(\cdot)$. It can be shown that (5.61) is applicable for any arbitrary vector phasors $\overline{E}$ and $\overline{H}$; the time-average power density carried by the fields $\overline{E}$ and $\overline{H}$ is

$$\overline{S}_{av} = \frac{1}{2} \text{Re} \left(\overline{E} \times \overline{H}^*\right). \tag{5.62}$$
$$time-average\ Poynting\ vector$$

5.5 Scalar and Vector Potentials

Maxwell's equations describe time-varying electromagnetic fields. One way of solving Maxwell's equations is in terms of a magnetic vector potential. The magnetic vector potential enables us to compute electromagnetic fields systematically. The concept of magnetic vector potential was previously introduced to solve magnetostatic problems, where fields were time-invariant. In this section, we will present the magnetic vector potential that is applicable even for time-varying fields. We will introduce the magnetic vector potential starting from Maxwell's equations for an isotropic and homogeneous medium (μ, ϵ: scalar and constant). To this end, we compare the magnetic Gauss's law $\nabla \cdot \overline{B} = 0$, with the vector identity $\nabla \cdot \left(\nabla \times \overline{A}\right) = 0$, to obtain

$$\overline{B} = \nabla \times \overline{A} \Longrightarrow \overline{H} = \frac{1}{\mu} \nabla \times \overline{A} \tag{5.63}$$

where $\overline{A}$ is called the magnetic vector potential. We substitute $\overline{B} = \nabla \times \overline{A}$ into Faraday's law

$$\nabla \times \overline{E} = -j\omega \overline{B} \tag{5.64}$$

to obtain

$$\nabla \times \left(\overline{E} + j\omega \overline{A}\right) = 0. \tag{5.65}$$

Since any scalar function V satisfies the relation

$$\nabla \times (-\nabla V) = 0 \tag{5.66}$$

we let

$$\overline{E} + j\omega\overline{A} = -\nabla V \qquad (5.67)$$

where V is called the electric scalar potential. Hence

$$\overline{E} = -j\omega\overline{A} - \nabla V. \qquad (5.68)$$

Then we substitute

$$\overline{H} = \frac{1}{\mu}\nabla \times \overline{A} \qquad (5.69)$$

$$\overline{E} = -j\omega\overline{A} - \nabla V \qquad (5.70)$$

into Ampère's law

$$\nabla \times \overline{H} = \overline{J} + j\omega\epsilon\overline{E} \qquad (5.71)$$

to obtain

$$-\nabla \times \nabla \times \overline{A} + \mu\epsilon\omega^2\overline{A} - j\omega\mu\epsilon\nabla V = -\mu\overline{J}. \qquad (5.72)$$

Since

$$\nabla \times \nabla \times \overline{A} = -\nabla^2\overline{A} + \nabla\left(\nabla \cdot \overline{A}\right) \qquad (5.73)$$

we rewrite (5.72) as

$$\nabla^2\overline{A} + \mu\epsilon\omega^2\overline{A} - \nabla\left[(\nabla \cdot \overline{A}) + j\omega\mu\epsilon V\right] = -\mu\overline{J}. \qquad (5.74)$$

On the other hand, the divergence of (5.68) gives

$$\nabla \cdot \overline{E} + j\omega\nabla \cdot \overline{A} = -\underbrace{\nabla \cdot \nabla V}_{\nabla^2 V}. \qquad (5.75)$$

Using Gauss's law $\nabla \cdot \overline{E} = \dfrac{\rho_v}{\epsilon}$, we obtain

$$\nabla^2 V + j\omega\nabla \cdot \overline{A} = -\frac{\rho_v}{\epsilon}. \qquad (5.76)$$

It is important to note that the magnetic vector potential $\overline{A}$ up to now is defined in terms of $\nabla \times \overline{A}$. Defining $\overline{A}$ in terms of $\nabla \times \overline{A}$ alone is insufficient for a unique determination of $\overline{A}$. The divergence condition $\nabla \cdot \overline{A}$ must be specified in addition to $\nabla \times \overline{A}$. To make $\overline{A}$ unique, we choose the Lorentz condition

$$\nabla \cdot \overline{A} = -j\omega\mu\epsilon V \qquad (5.77)$$

and simplify (5.74) and (5.76) to

$$\nabla^2 \overline{A} + k^2 \overline{A} = -\mu \overline{J} \tag{5.78}$$

$$\nabla^2 V + k^2 V = -\frac{\rho_v}{\epsilon} \tag{5.79}$$

Helmholtz's equations

where $k \ (= \omega \sqrt{\mu \epsilon})$ is called the wave number. The fields are given in terms of $\overline{A}$ as

$$\overline{E} = -j\omega \overline{A} - \frac{j}{\omega \mu \epsilon} \nabla (\nabla \cdot \overline{A}) \tag{5.80}$$

$$\overline{H} = \frac{1}{\mu} \nabla \times \overline{A}. \tag{5.81}$$

fields in terms of magnetic vector potential

In electromagnetic boundary-value problems, the importance of Helmholtz's equation for the magnetic vector potential cannot be overemphasized. The strategy here is first to solve Helmholtz's equation for $\overline{A}$ and then subsequently to evaluate $\overline{E}$ and $\overline{H}$ from $\overline{A}$. The magnetic vector potential is regarded as an intermediate step for evaluating electromagnetic fields, which are our ultimate solutions.

Static-limit case: In static limit (ω, $k \to 0$), Helmholtz's equations and fields are reduced to the results of electrostatics and magnetostatics.

1. For electrostatic fields

$$\nabla^2 V = -\frac{\rho_v}{\epsilon} \tag{5.82}$$

$$\overline{E} = -\nabla V. \tag{5.83}$$

2. For magnetostatic fields

$$\nabla^2 \overline{A} = -\mu \overline{J} \tag{5.84}$$

$$\overline{H} = \frac{1}{\mu} \nabla \times \overline{A}. \tag{5.85}$$

Fig. 5.6 Boundary between
dielectric medium 1 and
medium 2

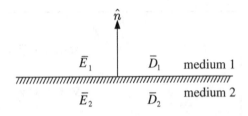

We note that (5.82) and (5.84) are Poisson's equations, which were previously
developed under the static assumption $\left(\dfrac{\partial}{\partial t} = 0\right)$.

5.6 Problems for Chapter 5

1. In Fig. 5.6, the boundary conditions for electrostatic fields $(\partial/\partial t = 0)$ are
   given by

$$\hat{n} \times \left(\overline{E}_1 - \overline{E}_2\right) = 0 \tag{5.86}$$

$$\hat{n} \cdot \left(\overline{D}_1 - \overline{D}_2\right) = \rho_s. \tag{5.87}$$

   Rewrite the above boundary conditions in terms of the electric scalar potential V
   where $\overline{E} = -\nabla V$.

2. Prove the relation $\overline{\mathcal{E}} \cdot \dfrac{\partial \overline{\mathcal{E}}}{\partial t} = \dfrac{1}{2}\dfrac{\partial |\overline{\mathcal{E}}|^2}{\partial t}$.

Chapter 6
Uniform Plane Waves

6.1 Waves in Lossless Media

Can electromagnetic fields really travel in space at the speed of light? The answer to this question comes from Maxwell's equations. In a source-free $\left(\overline{J} = 0 \text{ and } \rho_v = 0\right)$, homogeneous $(\mu, \epsilon: \text{constant})$, and lossless $(\sigma = 0)$ medium, Maxwell's equations are written as

$$\nabla \times \overline{E} = -j\omega\mu\overline{H} \tag{6.1}$$
$$\nabla \times \overline{H} = j\omega\epsilon\overline{E} \tag{6.2}$$
$$\nabla \cdot \overline{E} = 0 \tag{6.3}$$
$$\nabla \cdot \overline{H} = 0. \tag{6.4}$$

It is possible to solve (6.1) and (6.2) simultaneously for $\overline{E}$ and $\overline{H}$. To eliminate $\overline{H}$, we consider

$$\nabla \times \nabla \times \overline{E} = -j\omega\mu \underbrace{\nabla \times \overline{H}}_{j\omega\epsilon\overline{E}} = \omega^2\mu\epsilon\overline{E}. \tag{6.5}$$

The vector relation shows

$$\nabla \times \nabla \times \overline{E} = -\nabla^2\overline{E} + \nabla\underbrace{\left(\nabla \cdot \overline{E}\right)}_{0}. \tag{6.6}$$

Equating (6.5) with (6.6) yields Helmholtz's equation

$$\nabla^2\overline{E} + k^2\overline{E} = 0 \Longrightarrow \begin{cases} \nabla^2 E_x + k^2 E_x = 0 \\[2mm] \nabla^2 E_y + k^2 E_y = 0 \\[2mm] \nabla^2 E_z + k^2 E_z = 0 \end{cases} \tag{6.7}$$

H. J. Eom, *Primary Theory of Electromagnetics*, Power Systems,
DOI: 10.1007/978-94-007-7143-7_6, © Springer Science+Business Media Dordrecht 2013

where $k = \omega\sqrt{\mu\epsilon}$ is called the wave number. This Helmholtz's equation describes electric fields in a source-free region. Alternatively, (6.7) can be derived from Helmholtz's equations for the potentials

$$\nabla^2 V + k^2 V = 0 \tag{6.8}$$
$$\nabla^2 \overline{A} + k^2 \overline{A} = 0. \tag{6.9}$$

Substituting the relation

$$\overline{A} = \frac{j}{\omega}(\overline{E} + \nabla V) \tag{6.10}$$

into (6.9) yields (6.7), as was expected. Next we investigate a field $\overline{E} = \hat{x} E_x$ where E_x is a function of z. Then Helmholtz's equation for E_x

$$\frac{d^2 E_x}{dz^2} + k^2 E_x = 0 \tag{6.11}$$

yields a solution

$$E_x = \underbrace{c_1 e^{-jkz}}_{E^+} + \underbrace{c_2 e^{jkz}}_{E^-} . \tag{6.12}$$

The nature of E^+ becomes more apparent in a time-varying form $\mathcal{E}^+$. Assuming $c_1 = |c_1| e^{j\phi}$, we consider the time-varying form

$$
\begin{aligned}
\mathcal{E}^+ &= \mathrm{Re}\left(E^+ e^{j\omega t}\right) \\
&= \mathrm{Re}\left(c_1 e^{-jkz+j\omega t}\right) \\
&= |c_1| \cos\left(kz - \omega t - \phi\right) \\
&= |c_1| \cos\left[k\left(z - \frac{\omega}{k}t\right) - \phi\right]
\end{aligned}
\tag{6.13}
$$

where its temporal behavior is illustrated in Fig. 6.1. The field $\mathcal{E}^+$ is seen to propagate along the $+z$-direction with a velocity u as

$$u = \frac{\omega}{k} = \frac{1}{\sqrt{\mu\epsilon}}. \tag{6.14}$$

In air, $u = \dfrac{1}{\sqrt{\mu_0\epsilon_0}} \approx 3 \times 10^8$ (m/s). The wavelength λ and the period T of wave are given by

Fig. 6.1 A wave traveling in the z-direction with a velocity u

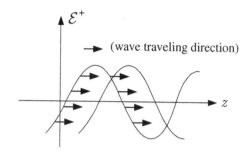

$$kλ = 2π \implies λ = \frac{2π}{k} \qquad (6.15)$$

$$ωT = 2π \implies T = \frac{2π}{ω}. \qquad (6.16)$$

Since the frequency is $f = 1/T$, we have

$$fλ = \frac{1}{T}\frac{2π}{k} = u. \qquad (6.17)$$

The magnetic field is straightforwardly obtained by substituting $\overline{E} = \hat{x}E^+$ into (6.1) as

$$\overline{H} = \hat{y}\underbrace{\frac{E^+}{η}}_{H^+} \qquad (6.18)$$

where $η = \sqrt{μ/ε}$ is a medium intrinsic impedance, which is about $120π$ ($Ω$) in air. The magnetic field also travels with a speed u. The wave of E^+ and H^+, propagating in the z-direction with a constant phase front on the x-y plane, is called a uniform plane wave. Similarly E^- and H^- propagate along the $-z$-direction with a velocity u as

$$\overline{E} = \hat{x}E^- = \hat{x}c_2 e^{jkz} \qquad (6.19)$$

$$\overline{H} = \hat{y}H^- = -\hat{y}\frac{E^-}{η}. \qquad (6.20)$$

General representation of uniform plane waves: Next we derive uniform plane waves that propagate in any arbitrary direction. Since $\overline{E}$ is a function of x, y, and z, a general solution to Helmholtz's equation (6.7) must be obtained. We first consider Helmholtz's equation for E_x as

$$\frac{\partial^2 E_x}{\partial x^2} + \frac{\partial^2 E_x}{\partial y^2} + \frac{\partial^2 E_x}{\partial z^2} + k^2 E_x = 0. \qquad (6.21)$$

We assume the solution in the product form of

$$E_x = XYZ \tag{6.22}$$

where X, Y, and Z are functions of x, y, and z, respectively. Substituting (6.22) into (6.21) and dividing (6.21) by XYZ, we obtain

$$\underbrace{\frac{1}{X}\frac{d^2X}{dx^2}}_{\text{function of } x} + \underbrace{\frac{1}{Y}\frac{d^2Y}{dy^2}}_{\text{function of } y} + \underbrace{\frac{1}{Z}\frac{d^2Z}{dz^2}}_{\text{function of } z} + k^2 = 0. \tag{6.23}$$

The first, second, and third terms in (6.23) must be equal to the constants $-k_x^2$, $-k_y^2$, and $-k_z^2$ irrespective of x, y, z as

$$\frac{1}{X}\frac{d^2X}{dx^2} = -k_x^2 \tag{6.24}$$

$$\frac{1}{Y}\frac{d^2Y}{dy^2} = -k_y^2 \tag{6.25}$$

$$\frac{1}{Z}\frac{d^2Z}{dz^2} = -k_z^2 \tag{6.26}$$

where

$$k_x^2 + k_y^2 + k_z^2 = k^2. \tag{6.27}$$

We choose the solutions as

$$X = c_1 e^{-jk_x x} \tag{6.28}$$
$$Y = c_2 e^{-jk_y y} \tag{6.29}$$
$$Z = c_3 e^{-jk_z z} \tag{6.30}$$

leading to

$$E_x = c_4 \exp\left[-j(k_x x + k_y y + k_z z)\right] \tag{6.31}$$

where c_4 is a constant amplitude. A similar discussion is possible for E_y and E_z. A general solution to Helmholtz's equation (6.7) is therefore

$$\overline{E} = \overline{E}_0 \exp\left[-j(k_x x + k_y y + k_z z)\right] \tag{6.32}$$

where $\overline{E}_0$ is a constant amplitude vector, which needs to be determined later. If k_x, k_y, and k_z are positive real numbers, (6.32) represents a uniform plane wave propagating along the positive x, y, and z axes, as illustrated in Fig. 6.2. It is convenient to introduce a wave vector

Fig. 6.2 A wave vector $\overline{k}$

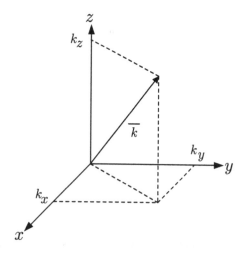

$$\overline{k} = \hat{x}k_x + \hat{y}k_y + \hat{z}k_z \tag{6.33}$$

and rewrite $\overline{E}$ as

$$\overline{E} = \overline{E}_0 \exp\left(-j\overline{k} \cdot \overline{r}\right) \tag{6.34}$$

where $\overline{r} = \hat{x}x + \hat{y}y + \hat{z}z$. Expression (6.34) represents a uniform plane wave propagating in the $\overline{k}$-direction, as shown in Fig. 6.2. Substituting (6.34) into (6.3) yields

$$\nabla \cdot \overline{E} = -j\overline{k} \cdot \underbrace{\overline{E}_0 \exp\left(-j\overline{k} \cdot \overline{r}\right)}_{\overline{E}} = 0 \tag{6.35}$$

implying that the wave vector $\overline{k}$ and the electric field vector $\overline{E}$ are perpendicular to each other. Furthermore we have

$$\overline{H} = \frac{j}{\omega\mu} \nabla \times \overline{E} = \frac{1}{\omega\mu}\overline{k} \times \overline{E} \tag{6.36}$$

showing that $\overline{E}$, $\overline{H}$, and $\overline{k}$ are mutually perpendicular. A graphical representation of a uniform plane wave in terms of $\overline{k}$, $\overline{E}$, and $\overline{H}$ is shown in Fig. 6.3. A uniform plane wave is a transverse electromagnetic (TEM) wave since electromagnetic field components are all transverse to the wave propagation direction $\overline{k}$. The time-average Poynting vector $\overline{S}_{av}$ of a uniform plane wave is

$$\overline{S}_{av} = \frac{1}{2} \operatorname{Re}\left(\overline{E} \times \overline{H}^*\right)$$
$$= \frac{1}{2}\left(\overline{E}_0 \cdot \overline{E}_0^*\right) \frac{\overline{k}}{\omega\mu} \tag{6.37}$$

Fig. 6.3 A uniform plane
wave $\overline{E}$ and $\overline{H}$ propagating in
the $\overline{k}$ direction

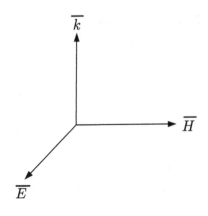

which shows that real power flows in the $\overline{k}$-direction.

6.2 Waves in Conductive Media

When waves propagate in conductive media, they become attenuated due to medium
losses. Although medium losses are rather complicated to categorize in general, we
will assume that medium losses are solely attributed to a medium conductivity σ.
In a source-free, homogeneous, and conductive medium, waves obey the following
Maxwell's equations:

$$\nabla \times \overline{E} = -j\omega\mu\overline{H} \tag{6.38}$$

$$\nabla \times \overline{H} = j\omega\epsilon\overline{E} + \sigma\overline{E} = j\omega\underbrace{\epsilon\left(1 - j\frac{\sigma}{\omega\epsilon}\right)}_{\epsilon_c}\overline{E} \tag{6.39}$$

$$\nabla \cdot \overline{E} = 0 \tag{6.40}$$

$$\nabla \cdot \overline{H} = 0 \tag{6.41}$$

where ϵ_c is a complex permittivity of conductive medium. From (6.38) and (6.39),
we obtain Helmholtz's equation for conductive medium

$$\nabla^2\overline{E} + k_c^2\overline{E} = 0 \tag{6.42}$$

which yields

$$\frac{d^2 E_x}{dz^2} + k_c^2 E_x = 0 \tag{6.43}$$

where $k_c = \omega\sqrt{\mu\epsilon_c}$. We write a solution to (6.43) as

$$E_x = c_1 e^{-jk_c z}.$$

(6.44)

Let α and β denote the real and imaginary parts of $jk_c \left[= j\omega\sqrt{\mu\epsilon\left(1 - j\dfrac{\sigma}{\omega\epsilon}\right)} \right]$ as

$$jk_c = \alpha + j\beta$$

(6.45)

where α and β are an attenuation constant and a phase constant, respectively. In order to evaluate α and β, we let

$$1 - j\frac{\sigma}{\omega\epsilon} = r\exp(-j\theta)$$

(6.46)

where $0 < \theta < \pi/2$ and

$$r = \sqrt{1 + \left(\frac{\sigma}{\omega\epsilon}\right)^2}$$

(6.47)

$$\theta = \tan^{-1}\left(\frac{\sigma}{\omega\epsilon}\right).$$

(6.48)

Thus

$$\sqrt{1 - j\frac{\sigma}{\omega\epsilon}} = \sqrt{r}\exp(-j\theta/2) = \sqrt{\frac{r+1}{2}} - j\sqrt{\frac{r-1}{2}}.$$

(6.49)

Final results are

$$\alpha = \omega\sqrt{\mu\epsilon}\sqrt{\frac{r-1}{2}}$$

(6.50)

$$\beta = \omega\sqrt{\mu\epsilon}\sqrt{\frac{r+1}{2}}.$$

(6.51)

When $c_1 = |c_1|e^{j\phi}$, the time-varying form is

$$\mathcal{E}_x = \text{Re}\left(c_1 e^{-jk_c z + j\omega t}\right)$$

$$= \text{Re}\left[c_1 e^{-(\alpha + j\beta)z + j\omega t}\right]$$

$$= |c_1|e^{-\alpha z}\cos(\beta z - \omega t - \phi).$$

(6.52)

Figure 6.4 depicts the behavior of (6.52) versus the distance z at a certain instant t. It is seen that the field becomes exponentially decaying as it propagates along the z-direction. The wave attenuation is controlled by the attenuation constant α while the wave velocity is determined by the phase constant β. The constants α and β

Fig. 6.4 An attenuated wave
traveling in the z-direction

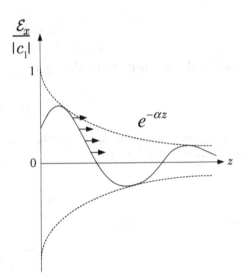

can be further simplified, depending on the size of conductivity σ. We consider two
different conductive media: a good dielectric and a good conducting medium.

1. If $\dfrac{\sigma}{\omega\epsilon} \ll 1$, the medium is a good dielectric and

$$r = \sqrt{1 + \left(\frac{\sigma}{\omega\epsilon}\right)^2} \approx 1 + \frac{1}{2}\left(\frac{\sigma}{\omega\epsilon}\right)^2 \tag{6.53}$$

$$\alpha \approx \frac{\sigma}{2}\sqrt{\frac{\mu}{\epsilon}} \tag{6.54}$$

$$\beta \approx \omega\sqrt{\mu\epsilon}. \tag{6.55}$$

2. If $\dfrac{\sigma}{\omega\epsilon} \gg 1$, the medium is a good conductor and

$$r = \sqrt{1 + \left(\frac{\sigma}{\omega\epsilon}\right)^2} \approx \frac{\sigma}{\omega\epsilon} \tag{6.56}$$

$$\alpha \approx \beta \approx \underbrace{\sqrt{\frac{\omega\mu\sigma}{2}}}_{1/\delta} \tag{6.57}$$

where δ is called the skin depth of a conductor. A wave within a good conductor
is written as

$$E_x = c_1 e^{-(1+j)z/\delta} \tag{6.58}$$

which represents an attenuated wave whose rate of attenuation is determined by the factor $\exp(-z/\delta)$. We note that the amount of field attenuation over the skin depth $z = \delta$ is $1/e \approx 0.368$. The skin depth of copper $(\sigma = 5.80 \times 10^7 \text{ (S/m)})$ at 100 (MHz) is $\delta = 6.6 \times 10^{-3}$ (mm), which is vanishingly small compared to the wavelength. Hence the field inside copper is approximately null at 100 (MHz). This implies that copper or any other good conductors at 100 (MHz) and even higher frequencies are practically perfect electric conductors of infinite conductivity.

6.3 Polarization of Plane Waves

The tip of the electric field vector, as the wave propagates, traces a certain curve on a plane perpendicular to the wave propagation direction. The trace of the tip of the electric field vector is explained in terms of the polarization of plane wave. To understand the polarization of plane wave, we will consider a uniform plane wave propagating along the z-direction as

$$\overline{E} = \left[\hat{x} E_1 \exp\left(j\theta_x\right) + \hat{y} E_2 \exp\left(j\theta_y\right)\right] e^{-jkz} \tag{6.59}$$

where E_1 and E_2 are the magnitudes of $\overline{E}$ in the x- and y-directions, respectively, and θ_x and θ_y are the corresponding phases. It is convenient to explain the polarization of plane wave by using its time-varying form. The time-varying form of $\overline{E}$ is expressed as

$$\overline{\mathcal{E}} = \text{Re}\left(\overline{E} e^{j\omega t}\right) = \hat{x}\,\mathcal{E}_x + \hat{y}\,\mathcal{E}_y \tag{6.60}$$

where

$$\mathcal{E}_x = E_1 \cos\left(\omega t - kz + \theta_x\right) \tag{6.61}$$
$$\mathcal{E}_y = E_2 \cos\left(\omega t - kz + \theta_y\right). \tag{6.62}$$

1. **Linear Polarization** When $\theta_y - \theta_x = 0$ or π, the ratio of $\mathcal{E}_y$ and $\mathcal{E}_x$ becomes

$$\frac{\mathcal{E}_y}{\mathcal{E}_x} = \pm\frac{E_2}{E_1} \tag{6.63}$$

where the tip of the electric field traces a straight line on the $\mathcal{E}_x$-$\mathcal{E}_y$ plane and the wave is said to have linear polarization. Its graphical representation is shown in Fig. 6.5.

2. **Circular Polarization** When $\theta_y - \theta_x = \pm\pi/2$ and $E_1 = E_2$, the field becomes

Fig. 6.5 Linear polarization

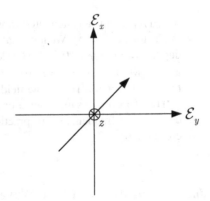

$$\mathcal{E}_x = E_1 \cos{(\omega t - kz + \theta_x)} \tag{6.64}$$
$$\mathcal{E}_y = \mp E_1 \sin{(\omega t - kz + \theta_x)}. \tag{6.65}$$

The trace of the field is given by

$$\left(\frac{\mathcal{E}_x}{E_1}\right)^2 + \left(\frac{\mathcal{E}_y}{E_1}\right)^2 = 1 \tag{6.66}$$

$$\tan^{-1}\left(\frac{\mathcal{E}_y}{\mathcal{E}_x}\right) = \mp(\omega t - kz + \theta_x) = \psi \tag{6.67}$$

which represents a circle on the $\mathcal{E}_x$-$\mathcal{E}_y$ plane and the wave is said to have circular polarization. When $\theta_y - \theta_x = \pi/2$, then

$$\psi = -(\omega t - kz + \theta_x) \Longrightarrow \frac{d\psi}{dt} = -\omega < 0 \,. \tag{6.68}$$

The electric field vector behaves as a left-handed screw advancing in the $+z$-direction and this polarization is called left-hand circular polarization. The tip of electric field vector is seen to rotate in the counterclockwise direction, as illustrated in Fig. 6.6. Similarly, when $\theta_y - \theta_x = -\pi/2$, then

$$\psi = \omega t - kz + \theta_x \Longrightarrow \frac{d\psi}{dt} = \omega > 0 \tag{6.69}$$

and the electric field vector behaves as a right-handed screw advancing in the $+z$-direction. This polarization is called right-hand circular polarization. Its graphical representation is given in Fig. 6.7.

Fig. 6.6 Left-hand circular polarization when the wave propagates in the z-direction (into the page)

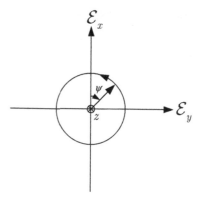

Fig. 6.7 Right-hand circular polarization when the wave propagates in the z-direction (into the page)

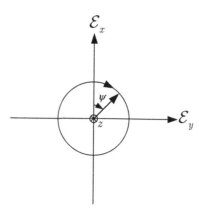

6.4 Reflection and Transmission

When a uniform plane wave impinges on a boundary between two different media, the wave is reflected and transmitted at the boundary. Reflected and transmitted waves are governed by Maxwell's equations. This problem of wave reflection and transmission can be solved by enforcing the boundary conditions at the boundary. Next we will investigate wave reflection and transmission at a planar interface between two lossless media.

6.4.1 Perpendicular Polarization

Figure 6.8 illustrates the wave reflection and transmission at the boundary between two lossless media whose permittivities and permeabilities are all real numbers. Since the electric field vector (y-component) is perpendicular to the plane of incidence (x-z plane), this case is referred to as the perpendicular polarization. The wave

Fig. 6.8 Reflection and transmission of a wave across a planar boundary: perpendicular polarization

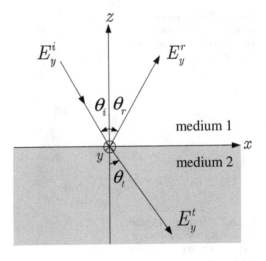

Fig. 6.9 A wave vector $\bar{k}$ for the incident wave

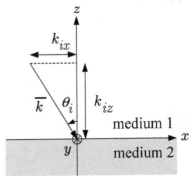

numbers in medium 1 and medium 2 are k_1 ($= \omega\sqrt{\mu_1\epsilon_1}$) and k_2 ($= \omega\sqrt{\mu_2\epsilon_2}$), respectively. Figure 6.9 illustrates a wave vector $\bar{k}$ of the incident wave with an incident angle θ_i. The incident uniform plane wave is

$$E_y^i = E_0\, e^{-j\bar{k}\cdot\bar{r}} \tag{6.70}$$

where

$$\bar{k} = \hat{x}\, \underbrace{k_1 \sin\theta_i}_{k_{ix}} -\hat{z}\, \underbrace{k_1 \cos\theta_i}_{k_{iz}} \tag{6.71}$$

$$\bar{r} = \hat{x}x + \hat{y}y + \hat{z}z. \tag{6.72}$$

Hence

$$E_y^i = E_0 \exp(-jk_{ix}x + jk_{iz}z). \tag{6.73}$$

The reflected electric field is

$$E_y^r = RE_0 \exp(-jk_x'x - jk_z'z) \tag{6.74}$$

where R is the reflection coefficient, $k_x' = k_1 \sin \theta_r$, and $k_z' = k_1 \cos \theta_r$. The transmitted wave in medium 2 is expressed as

$$E_y^t = TE_0 \exp(-jk_x''x + jk_z''z) \tag{6.75}$$

where $k_x'' = k_2 \sin \theta_t$, $k_z'' = k_2 \cos \theta_t$, and T is the transmission coefficient. Until now, we have introduced four unknown parameters, θ_r, θ_t, R, and T, which can be determined by imposing the boundary conditions. The boundary condition requires that the tangential electric field slightly above $z = 0$ should be the same as the tangential electric field slightly below $z = 0$. The tangential electric field continuity

$$E_y^i\big|_{z=0} + E_y^r\big|_{z=0} = E_y^t\big|_{z=0} \tag{6.76}$$

is rewritten as

$$\exp(-jk_{ix}x) + R\exp(-jk_x'x) = T\exp(-jk_x''x). \tag{6.77}$$

To satisfy (6.77) irrespective of x, we let

$$k_{ix} = k_x' = k_x'' \tag{6.78}$$
$$1 + R = T \tag{6.79}$$

subsequently yielding

$$\theta_i = \theta_r \tag{6.80}$$

and Snell's law

$$k_1 \sin \theta_i = k_2 \sin \theta_t. \tag{6.81}$$

Faraday's law $\nabla \times \overline{E} = -j\omega\mu\overline{H}$ yields

$$H_x = -\frac{j}{\omega\mu}\frac{\partial E_y}{\partial z}. \tag{6.82}$$

The tangential H_x continuity at $z = 0$

$$H_x^i\big|_{z=0} + H_x^r\big|_{z=0} = H_x^t\big|_{z=0} \tag{6.83}$$

gives

$$\frac{k_{iz}}{\mu_1}(R-1) = -\frac{k_z''}{\mu_2}T. \tag{6.84}$$

Solving (6.79) and (6.84) for R and T gives

$$R = \frac{Z_2 - Z_1}{Z_2 + Z_1} \tag{6.85}$$

$$T = \frac{2Z_2}{Z_2 + Z_1} \tag{6.86}$$

reflection and transmission coefficients for perpendicular polarization

where $Z_2 = \eta_2/\cos\theta_t$, $Z_1 = \eta_1/\cos\theta_i$, $\eta_2 = \sqrt{\mu_2/\epsilon_2}$, and $\eta_1 = \sqrt{\mu_1/\epsilon_1}$.

Example 6.1 Reflection from a perfectly conducting plane.

Consider reflection in Fig. 6.8, where medium 2 is perfectly conducting. The incident wave in medium 1 is $E_y^i = \exp(-jk_{ix}x + jk_{iz}z)$, where $k_{ix} = k_1 \sin\theta_i$ and $k_{iz} = k_1 \cos\theta_i$. Determine the electric field in medium 1 and the current density induced on the conducting surface.

Solution: The incident field cannot penetrate the perfectly conducting medium of zero skin depth; the transmitted wave in medium 2 is zero. The electric field reflected in medium 1 is

$$E_y^r = R\exp(-jk_x'x - jk_z'z) \tag{6.87}$$

where R is the reflection coefficient, $k_x' = k_1 \sin\theta_r$, and $k_z' = k_1 \cos\theta_r$. The tangential electric field continuity at the boundary $z = 0$

$$E_y^i\big|_{z=0} + E_y^r\big|_{z=0} = 0 \tag{6.88}$$

gives

$$\exp(-jk_{ix}x) + R\exp(-jk_x'x) = 0. \tag{6.89}$$

Hence we are led to

$$k_{ix} = k_x' \implies \theta_i = \theta_r \tag{6.90}$$

$$R = -1 \tag{6.91}$$

$$E_y^r = -\exp(-jk_{ix}x - jk_{iz}z). \tag{6.92}$$

The electric field in medium 1 is

Fig. 6.10 Reflection and transmission of a wave across a planar boundary: parallel polarization

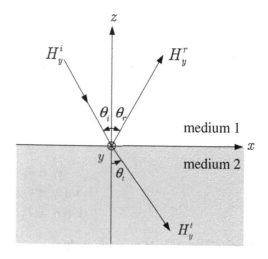

$$E_y = E_y^i + E_y^r = 2j \, \exp\left(-jk_{ix}x\right) \sin k_{iz}z. \tag{6.93}$$

The induced surface current density is

$$\overline{J}_s = \underbrace{\hat{n}}_{\hat{z}} \times \overline{H}\big|_{z=0}$$

$$= \hat{y} H_x\big|_{z=0}$$

$$= \hat{y} 2 \sqrt{\frac{\epsilon_1}{\mu_1}} \cos \theta_i \, \exp(-jk_{ix}x). \tag{6.94}$$

6.4.2 Parallel Polarization

Consider the wave reflection and transmission in Fig. 6.10, where the magnetic field vector has a y-component. Since the electric field vector is parallel to the plane of incidence, this case is called the parallel polarization. In medium 1 ($k_1 = \omega\sqrt{\mu_1\epsilon_1}$), the magnetic fields are

$$H_y^i = H_0 \exp(-jk_{ix}x + jk_{iz}z) \tag{6.95}$$

$$H_y^r = R H_0 \exp(-jk_x'x - jk_z'z). \tag{6.96}$$

In medium 2 ($k_2 = \omega\sqrt{\mu_2\epsilon_2}$), the magnetic field is

$$H_y^t = T H_0 \exp(-jk_x'' x + jk_z'' z) \tag{6.97}$$

where $k_{ix} = k_1 \sin\theta_i$, $k_{iz} = k_1 \cos\theta_i$, $k_x' = k_1 \sin\theta_r$, $k_x'' = k_2 \sin\theta_t$, $k_z' = k_1 \cos\theta_r$, and $k_z'' = k_2 \cos\theta_t$. The tangential magnetic field continuity at the boundary $z = 0$

$$H_y^i\Big|_{z=0} + H_y^r\Big|_{z=0} = H_y^t\Big|_{z=0} \tag{6.98}$$

yields

$$\theta_i = \theta_r \tag{6.99}$$
$$k_1 \sin\theta_i = k_2 \sin\theta_t \tag{6.100}$$
$$1 + R = T. \tag{6.101}$$

Ampère's law $\nabla \times \overline{H} = j\omega\epsilon\overline{E}$ produces

$$E_x = \frac{j}{\omega\epsilon}\frac{\partial H_y}{\partial z}. \tag{6.102}$$

The E_x continuity

$$E_x^i\Big|_{z=0} + E_x^r\Big|_{z=0} = E_x^t\Big|_{z=0} \tag{6.103}$$

yields

$$\frac{k_{iz}}{\epsilon_1}(R - 1) = -\frac{k_z''}{\epsilon_2}T. \tag{6.104}$$

Equations (6.101) and (6.104) give

$$R = \frac{Y_2 - Y_1}{Y_2 + Y_1} \tag{6.105}$$

$$T = \frac{2Y_2}{Y_2 + Y_1} \tag{6.106}$$

reflection and transmission coefficients for parallel polarization

---

where $Y_2 = 1/(\eta_2 \cos\theta_t)$, $Y_1 = 1/(\eta_1 \cos\theta_i)$, $\eta_2 = \sqrt{\mu_2/\epsilon_2}$, and $\eta_1 = \sqrt{\mu_1/\epsilon_1}$.

6.5 Problems for Chapter 6

1. Derive (6.49).

   Hint: Use the formula $\cos\theta = 2\cos^2\left(\dfrac{\theta}{2}\right) - 1 = 1 - 2\sin^2\left(\dfrac{\theta}{2}\right)$.

2. Consider a wave propagating in a good conductor where $\dfrac{\sigma}{\omega\epsilon} \gg 1$. Show that the phase difference between the electric and magnetic fields is approximately $\dfrac{\pi}{4}$ radians.

3. Show that a uniform plane wave of (6.59) can be decomposed into a right-hand circularly polarized wave and a left-hand circularly polarized wave.

4. Consider a wave obliquely incident on a dielectric half-space with an incident angle θ_i, as illustrated in Fig. 6.10. Determine θ_i for total transmission when $\mu_1 = \mu_2 = \mu_0$.

5. Figure 6.10 shows a wave incident on a planar boundary between medium 1 and medium 2. If medium 1 is denser than medium 2 ($\epsilon_1 > \epsilon_2$ and $\mu_1 = \mu_2 = \mu_0$), a total reflection occurs under certain circumstances. Derive the condition for the total internal reflection.

6. A right-hand circularly polarized wave is normally incident on a perfectly conducting plane, as shown in Fig. 6.11. Show that the reflected wave is left-hand circularly polarized.

   Hint: Assume the incident wave $\overline{E} = (\hat{x} + j\hat{y})e^{jkz}$ and use the boundary conditions to determine the reflected wave.

7. Figure 6.12 shows a wave $E_y^i = e^{jkz}$ normally incident on a lossless dielectric slab whose thickness d is given by $k_1 d = \pi$ (k_1: wave number in a dielectric slab). Evaluate the time-average power reflected at $z = 0$.

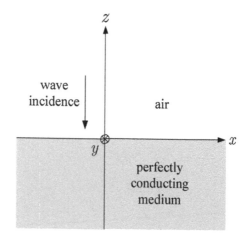

Fig. 6.11 A wave impinging on a perfectly conducting medium (PEC)

Fig. 6.12 A wave impinging
on a dielectric slab

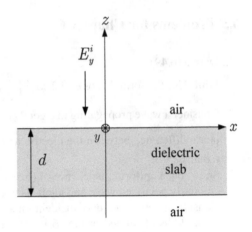

Chapter 7
Transmission Lines

7.1 Fundamentals of Transmission Lines: Field Approach

Figure 7.1 shows perfectly conducting parallel plates that guide electromagnetic waves along the z-direction. The interior of parallel plates is filled with a lossless material of permittivity ϵ and permeability μ. Assume that the spacing h between plates is small and the fringing field effect is ignored. The electric field within the plates has an x-component ($\overline{E} = \hat{x} E_x$), which is independent of x and y ($\partial E_x/\partial x = \partial E_x/\partial y = 0$). The field E_x obeys Helmholtz's equation

$$\frac{d^2 E_x}{dz^2} + k^2 E_x = 0 \tag{7.1}$$

where $k = \omega\sqrt{\mu\epsilon}$ is the wave number. The solution is

$$E_x = \underbrace{E_0^+ e^{-jkz}}_{E^+} + \underbrace{E_0^- e^{jkz}}_{E^-} . \tag{7.2}$$

Here E^+ represents a forward wave propagating in the $+z$-direction with a speed $u = 1/\sqrt{\mu\epsilon}$. Conversely E^- represents a backward wave propagating in the $-z$-direction. It is seen that the electric field E_x on the transmission line is the sum of forward and backward waves. Further boundary conditions would be required to determine the unknown coefficients E_0^+ and E_0^-. Faraday's law $\nabla \times \overline{E} = -j\omega\mu\overline{H}$ gives the magnetic field

H. J. Eom, *Primary Theory of Electromagnetics*, Power Systems, 147
DOI: 10.1007/978-94-007-7143-7_7, © Springer Science+Business Media Dordrecht 2013

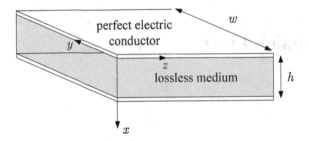

Fig. 7.1 Conducting parallel plates filled with a lossless medium

$$H_y = \underbrace{\left(\frac{E_0^+}{\eta}e^{-jkz}\right)}_{H^+} + \underbrace{\left(-\frac{E_0^-}{\eta}e^{jkz}\right)}_{H^-} \qquad (7.3)$$

where $\eta = \sqrt{\mu/\epsilon}$ is the intrinsic impedance of the medium. Electromagnetic waves can deliver power, which can be computed from the Poynting vector. The time-average power density carried by $\overline{E}$ and $\overline{H}$ is

$$\overline{S}_{av} = \frac{1}{2}\,\mathrm{Re}\left(\overline{E} \times \overline{H}^*\right) \qquad (7.4)$$

where the symbol $(\cdot)^*$ denotes a complex conjugate of $(\cdot)$. The time-average power delivery by E^+ and H^+ along the z-direction over a transverse cross section $(h \times w)$ is

$$
\begin{aligned}
P_{av} &= \int_0^w \int_0^h \overline{S}_{av} \cdot \hat{z}\,dx\,dy \\
&= \frac{1}{2}\mathrm{Re}\int_0^w \int_0^h E^+(H^+)^*\,dx\,dy \\
&= \frac{hw}{2\eta}|E_0^+|^2.
\end{aligned} \qquad (7.5)
$$

Although waves are composed of electric and magnetic fields, it is convenient to use voltages and currents for wave representation. The electric field of TEM wave on transmission lines can be written in terms of the electric scalar potential (or equivalently voltage). For the forward waves E^+ and H^+, we define a forward voltage V^+ as

$$V^+ = -\int_h^0 E^+\,dx = \underbrace{E_0^+ h}_{V_0^+}\,e^{-jkz} \qquad (7.6)$$

where the bottom plate at $x = h$ is taken as a reference for the voltage difference. To determine the current we use the boundary condition

$$\overline{J}_s = \hat{n} \times (\overline{H}_1 - \overline{H}_2) \tag{7.7}$$

where $\overline{J}_s$ is the surface current density on the top conducting plate and

$$\begin{cases} \overline{H}_1 = \hat{y} H^+ & \text{beneath } x = 0 \\ \overline{H}_2 = 0 & \text{above } x = 0. \end{cases} \tag{7.8}$$

Therefore $\overline{J}_s$ is

$$\overline{J}_s = \underbrace{\hat{n}}_{\hat{x}} \times \underbrace{\overline{H}_1}_{\hat{y} H^+} \Big|_{x=0} = \hat{z} H^+. \tag{7.9}$$

The forward current on the top plate is

$$I^+ = \int_0^w \overline{J}_s \cdot \hat{z} \, dy = \int_0^w H^+ \, dy = \underbrace{\frac{E_0^+ w}{\eta}}_{I_0^+} e^{-jkz}. \tag{7.10}$$

The forward waves V^+ and I^+ travel along the $+z$-direction and deliver the time-average power

$$P_{av} = \frac{1}{2} \text{Re} \left(V^+ I^{+*} \right) = \frac{hw}{2\eta} |E_0^+|^2. \tag{7.11}$$

The ratio of V^+ to I^+ is called the characteristic impedance Z_0 of the transmission line:

$$Z_0 = \frac{V^+}{I^+} = \sqrt{\frac{\mu}{\epsilon}} \frac{h}{w}. \tag{7.12}$$

The backward voltage V^- and current I^- are similarly defined as

$$V^- = -\int_h^0 E^- \, dx = \underbrace{E_0^- h \, e^{jkz}}_{V_0^-} \tag{7.13}$$

$$I^- = \int_0^w H^- \, dy = -\underbrace{\frac{E_0^- w}{\eta}}_{I_0^-} e^{jkz}. \tag{7.14}$$

Fig. 7.2 Voltage and current
representations on a transmission line

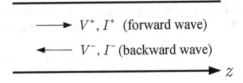

The ratio gives

$$\frac{V^-}{I^-} = -Z_0. \tag{7.15}$$

Graphical representations of forward and backward waves are shown in Fig. 7.2.
The total wave is the sum of forward and backward waves. The total voltage
$V \ (= E_x h)$ and the total current $I \ (= H_y w)$ are

$$
\begin{aligned}
V &= E_x h \\
&= E_0^+ h e^{-jkz} + E_0^- h e^{jkz} \\
&= \underbrace{V_0^+ e^{-jkz}}_{V^+} + \underbrace{V_0^- e^{jkz}}_{V^-}
\end{aligned} \tag{7.16}
$$

$$
\begin{aligned}
I &= H_y w \\
&= \frac{E_0^+}{\eta} w e^{-jkz} - \frac{E_0^-}{\eta} w e^{jkz} \\
&= \underbrace{\left(\frac{V_0^+}{Z_0} e^{-jkz} \right)}_{I^+} + \underbrace{\left(-\frac{V_0^-}{Z_0} e^{jkz} \right)}_{I^-}.
\end{aligned} \tag{7.17}
$$

Equations (7.16) and (7.17) represent a general solution to the transmission line problem, as illustrated in Fig. 7.2. The amplitudes V_0^+ and V_0^- should be determined by
imposing the boundary conditions on the receiving and sending ends of transmission
lines; relevant discussions are given in Sect. 7.3.

7.2 Fundamentals of Transmission Lines: Circuit Approach

A circuit approach is often used for transmission-line modeling due to simplicity.
The circuit approach is possible since the TEM wave on the cross section $(x - y$
plane) of a transmission line is governed by Laplace's equation for the electric scalar
potential. The electric scalar potential permits us to use capacitance and inductance
concepts, which were initially developed for static or slowly time-varying fields. In
what follows, we will show that a lossless transmission line can be modeled with

Fig. 7.3 A cross section
($h \times w$) of the parallel plates
carrying charges

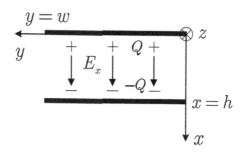

distributed shunt capacitances and series inductances. Figure 7.3 illustrates a cross section of parallel plates guiding waves along the $\pm z$-directions. A voltage V between two plates results in charge accumulation. The surface charge density at $x = 0$ is

$$\rho_s = \epsilon E_x. \tag{7.18}$$

A charge accumulated per unit length in the z-direction is

$$Q = \rho_s w = \epsilon E_x w. \tag{7.19}$$

Since $V = E_x h$, the capacitance per unit length is

$$C = \frac{Q}{V} = \frac{\epsilon w}{h}. \tag{7.20}$$

Figure 7.4 shows a cross-sectional view of the parallel plates carrying currents. The surface current density $\overline{J}_s$ at $x = 0$ is

$$\overline{J}_s = \hat{z} H_y \tag{7.21}$$

yielding the total current $I = w H_y$. Since the magnetic flux per unit length is

$$\Phi = \mu H_y h = \mu \frac{I}{w} h \tag{7.22}$$

Fig. 7.4 A cross section
($h \times w$) of the parallel plates
carrying a current I

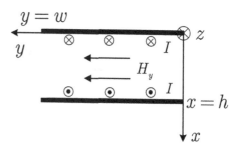

Fig. 7.5 Circuit representation of a transmission line segment dz

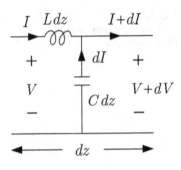

the inductance per unit length in the z-direction is

$$L = \frac{\Phi}{I} = \frac{\mu h}{w}. \tag{7.23}$$

Consider a differential length dz of transmission line in Figs. 7.3 and 7.4. Its equivalent circuit model is given in Fig. 7.5 in terms of the capacitance per unit length C and the inductance per unit length L. To obtain a relation between V and I, we consider a voltage across the inductor as

$$j\omega L\, dz I = -dV. \tag{7.24}$$

Hence

$$\frac{dV}{dz} = -j\omega L I. \tag{7.25}$$

The current through the capacitor is

$$j\omega C\, dz(V + dV) = -dI. \tag{7.26}$$

Since $V + dV \approx V$,

$$\frac{dI}{dz} = -j\omega C V. \tag{7.27}$$

Eliminating I from (7.25) and (7.27), we obtain

$$\frac{d^2 V}{dz^2} + \omega^2 LCV = 0. \tag{7.28}$$

Fig. 7.6 A coaxial cable

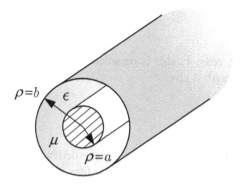

Since $\omega^2 LC = \omega^2 \mu\epsilon = k^2$, the solution is given by

$$V = \underbrace{V_0^+ e^{-jkz}}_{V^+} + \underbrace{V_0^- e^{jkz}}_{V^-}. \tag{7.29}$$

This shows that voltages propagate along the $\pm z$-directions with the wave number k. The corresponding current is

$$I = \frac{j}{\omega L}\frac{dV}{dz} = \underbrace{\sqrt{\frac{C}{L}}}_{\frac{1}{Z_0}}\left(V_0^+ e^{-jkz} - V_0^- e^{jkz}\right)$$

$$= \underbrace{\left(\frac{V_0^+}{Z_0}e^{-jkz}\right)}_{I^+} + \underbrace{\left(-\frac{V_0^-}{Z_0}e^{jkz}\right)}_{I^-}. \tag{7.30}$$

The characteristic impedance is expressed in terms of L and C as

$$Z_0 = \sqrt{\frac{L}{C}} = \sqrt{\frac{\mu}{\epsilon}}\frac{h}{w}. \tag{7.31}$$

It is seen that the circuit-approach results indeed agree with the field-approach results.

Example 7.1 Characteristic impedance of a coaxial cable.
Derive the characteristic impedance of a coaxial cable whose inner and outer radii are a and b, respectively, in Fig. 7.6. The annular space between perfect conductors is filled with a dielectric of permittivity ϵ and permeability $\mu = \mu_0$.

Solution: The characteristic impedance in terms of the capacitance per unit length C and the inductance per unit length L is

$$Z_0 = \sqrt{\frac{L}{C}}. \tag{7.32}$$

A coaxial cable is regarded as a cylindrical capacitor whose capacitance per unit length is given by

$$C = \frac{2\pi\epsilon}{\ln\left(\frac{b}{a}\right)}. \tag{7.33}$$

Refer to Example 2.6 for the derivation of C. Since the conductors are perfectly conducting and the current flow is practically confined on the conductor surfaces ($\rho = a$ and b), the magnetic field exists only within the annular region of $a < \rho < b$: the inductance per unit length is given as

$$L = \frac{\mu}{2\pi} \ln\left(\frac{b}{a}\right). \tag{7.34}$$

Refer to Example 4.3 for the detailed derivation of L. Hence

$$Z_0 = \sqrt{\frac{L}{C}} = \frac{1}{2\pi}\sqrt{\frac{\mu}{\epsilon}} \ln\left(\frac{b}{a}\right). \tag{7.35}$$

7.3 Terminated Transmission Lines

Practical transmission lines are finite in length. Figure 7.7 shows a lossless transmission line having an AC voltage source V_s at $z = 0$ with an internal impedance Z_s. The transmission line, terminated at $z = l$ with a load Z_l, is in AC steady state. Figure 7.8 illustrates a wave incidence on the load. When a wave (V^+, I^+) is incident on the load at $z = l$, the reflection occurs to generate a reflected wave (V^-, I^-). Since the incident and reflected waves simultaneously exist on the transmission line, the total voltage V and the total current I are given by

$$\begin{aligned} V &= V^+ + V^- \\ &= V_0^+ e^{-jkz} + V_0^- e^{jkz} \end{aligned} \tag{7.36}$$

$$\begin{aligned} I &= I^+ + I^- \\ &= \frac{1}{Z_0}\left(V_0^+ e^{-jkz} - V_0^- e^{jkz}\right) \end{aligned} \tag{7.37}$$

where V_0^+ and V_0^- are two unknown coefficients. Two boundary conditions at $z = 0$ and $z = l$ are required to determine V_0^+ and V_0^-. These boundary conditions are related to Kirchhoff's voltage laws. Kirchhoff's voltage law at $z = l$ is written as

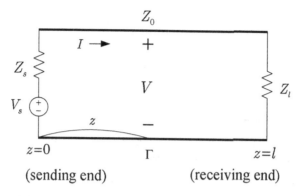

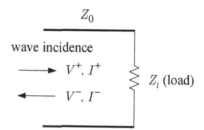

Fig. 7.7 A transmission line terminated with a load Z_l

Fig. 7.8 Incident and
reflected waves

$$Z_l = \frac{V}{I}\bigg|_{z=l} = Z_0 \frac{V_0^+ e^{-jkl} + V_0^- e^{jkl}}{V_0^+ e^{-jkl} - V_0^- e^{jkl}}. \tag{7.38}$$

Similarly Kirchhoff's voltage law at $z = 0$ is

$$V_s = Z_s I\big|_{z=0} + V\big|_{z=0}$$
$$= \frac{Z_s}{Z_0}\left(V_0^+ - V_0^-\right) + \left(V_0^+ + V_0^-\right). \tag{7.39}$$

Solving (7.38) and (7.39) for V_0^+ and V_0^-, we obtain

$$V_0^+ = \frac{Z_0}{(Z_0 + Z_s) + (Z_0 - Z_s)\Gamma_l e^{-2jkl}} V_s \tag{7.40}$$

$$V_0^- = \Gamma_l e^{-2jkl} V_0^+ \tag{7.41}$$

where $\Gamma_l = (Z_l - Z_0)/(Z_l + Z_0)$. Expressions (7.40) and (7.41) present a formal solution to the problem illustrated in Fig. 7.7. Voltage reflection coefficients
and impedances are often used to analyze transmission lines. The voltage reflection
coefficient Γ, a ratio of V^- to V^+, is expressed as

$$\Gamma = \frac{V^-}{V^+} = \underbrace{\frac{Z_l - Z_0}{Z_l + Z_0}}_{\Gamma_l} e^{-2jk(l-z)}. \qquad (7.42)$$

voltage reflection coefficient

Here Γ_l is seen to be the voltage reflection coefficient at the load ($z = l$). The impedance Z, a ratio of V to I, is given as

$$Z = \frac{V}{I} = Z_0 \frac{1+\Gamma}{1-\Gamma}. \qquad (7.43)$$

impedance

VSWR: Next we will study wave patterns that are generated on transmission lines due to the combination of forward and backward waves. To investigate voltages versus z, we consider the total voltage

$$V = V_0^+ e^{-jkz} + V_0^- e^{jkz}. \qquad (7.44)$$

Assuming $V_0^\pm = |V_0^\pm| \exp(j\theta^\pm)$, we convert V into a time-varying form as

$$\mathcal{V} = \underbrace{|V_0^+| \cos(\omega t + \theta^+ - kz) +}_{\mathcal{V}^+}$$
$$\underbrace{|V_0^-| \cos(\omega t + \theta^- + kz)}_{\mathcal{V}^-}. \qquad (7.45)$$

As the voltages $\mathcal{V}^+$ and $\mathcal{V}^-$ propagate oppositely in the $\pm z$-directions, their sum generates the standing wave $\mathcal{V}$ with alternating maxima and minima. Maximum $|\mathcal{V}|$ occurs when the crests of $\mathcal{V}^+$ and $\mathcal{V}^-$ coincide, as shown in Fig. 7.9a. This occurs when the voltages $\mathcal{V}^+$ and $\mathcal{V}^-$ are in phase, yielding

$$|\mathcal{V}|_{max} = |V_0^+| + |V_0^-|. \qquad (7.46)$$

After maximum occurrence, each wave travels over a distance $k\Delta z = \pi/2$ to generate minimum $|\mathcal{V}|$, as shown in Fig. 7.9b, where the voltages $\mathcal{V}^+$ and $\mathcal{V}^-$ are out of phase. Hence

$$|\mathcal{V}|_{min} = |V_0^+| - |V_0^-|. \qquad (7.47)$$

The distance Δz between adjacent minimum and maximum locations is

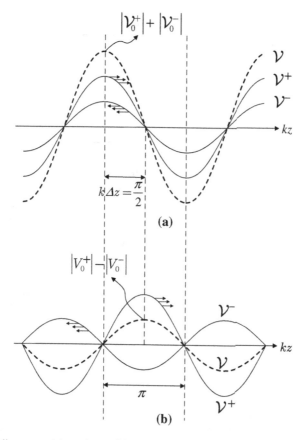

Fig. 7.9 Standing waves (**a**) maximum (**b**) minimum

$$k\Delta z = \frac{\pi}{2} \implies \Delta z = \frac{\lambda}{4}. \tag{7.48}$$

The voltage standing-wave ratio (VSWR) is often used to quantify the wave reflection on transmission lines. It is expressed as

$$
\begin{aligned}
\text{VSWR} &= \frac{|\mathcal{V}|_{max}}{|\mathcal{V}|_{min}} \\
&= \frac{|V_0^+| + |V_0^-|}{|V_0^+| - |V_0^-|} \\
&= \frac{1 + |\Gamma|}{1 - |\Gamma|}.
\end{aligned} \tag{7.49}
$$

Fig. 7.10 (a) Unmatched
transmission line and (b)
matched transmission line

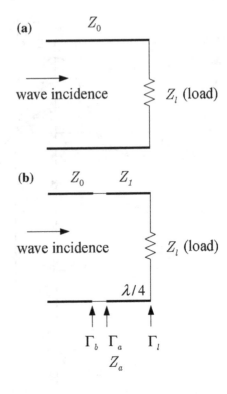

Example 7.2 A quarter-wave transformer.

Figure 7.10a shows a lossless transmission line of characteristic impedance Z_0 terminated with a resistive load $Z_l \neq Z_0$. Reflection exists due to the impedance mismatch between the load and the line. To eliminate this reflection, a quarter-wave transformer ($\lambda/4$ line, λ: wavelength) is often used. Figure 7.10b shows a quarter-wave transformer of the characteristic impedance Z_1 inserted between Z_l and Z_0. Determine Z_1.

Solution: It is convenient to compute reflection coefficients and impedances successively starting from the load. The voltage reflection coefficient Γ_l at the load is

$$\Gamma_l = \frac{Z_l - Z_1}{Z_l + Z_1}. \tag{7.50}$$

The voltage reflection coefficient Γ_a at the left end of the $\lambda/4$ line is

$$\Gamma_a = \Gamma_l e^{-2jk\lambda/4} = -\frac{Z_l - Z_1}{Z_l + Z_1}. \tag{7.51}$$

The impedance Z_a at the left end of the $\lambda/4$ line is

$$Z_a = Z_1 \frac{1 + \Gamma_a}{1 - \Gamma_a} = Z_1^2 / Z_l. \tag{7.52}$$

The voltage reflection coefficient Γ_b at the end of the line (Z_0) is

$$\Gamma_b = \frac{Z_a - Z_0}{Z_a + Z_0}. \tag{7.53}$$

The reflection disappears $(\Gamma_b = 0)$ if we choose

$$Z_0 = Z_a \implies Z_1 = \sqrt{Z_0 Z_l}. \tag{7.54}$$

Hence, the transmission line is matched to the load by using the quarter-wave transformer with $Z_1 = \sqrt{Z_0 Z_l}$.

7.4 Smith Chart

The Smith chart is a useful graphical tool for estimating impedances, reflection coefficients, etc. in microwave circuit and antenna applications. The Smith chart can be constructed from the relation of impedance and reflection coefficient. The impedance Z and the voltage reflection coefficient Γ on a transmission line with a characteristic impedance Z_0 are related by

$$\frac{Z}{Z_0} = \frac{1 + \Gamma}{1 - \Gamma}. \tag{7.55}$$

The Smith chart is based on a graphical description of (7.55). To construct the Smith chart, we separate $\frac{Z}{Z_0}$ and Γ into real and imaginary parts by letting

$$\frac{Z}{Z_0} = z = r + jx \tag{7.56}$$

$$\Gamma = \Gamma_r + j\Gamma_i. \tag{7.57}$$

Here z, r, and x are the normalized impedance, the normalized resistance, and the normalized reactance, respectively; Γ_r and Γ_i are the real and imaginary parts of Γ, respectively. Substituting (7.56) and (7.57) into (7.55) and rearranging the expressions, we obtain

Fig. 7.11 Smith chart

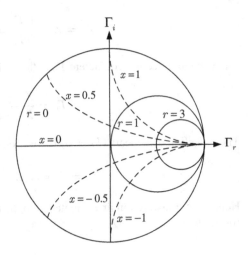

$$r = \frac{1 - \Gamma_r^2 - \Gamma_i^2}{(1 - \Gamma_r)^2 + \Gamma_i^2} \tag{7.58}$$

$$x = \frac{2\Gamma_i}{(1 - \Gamma_r)^2 + \Gamma_i^2}. \tag{7.59}$$

Rewriting (7.58) yields

$$\left(\Gamma_r - \frac{r}{1+r}\right)^2 + \Gamma_i^2 = \left(\frac{1}{1+r}\right)^2. \tag{7.60}$$

This equation represents a circle centered at $[\Gamma_r = r/(1+r),\ \Gamma_i = 0]$ with a radius $1/(1+r)$ on the $\Gamma_r - \Gamma_i$ plane. A family of curves (7.60) is plotted in solid lines in Fig. 7.11. Similarly (7.59) is rewritten as

$$(\Gamma_r - 1)^2 + \left(\Gamma_i - \frac{1}{x}\right)^2 = \left(\frac{1}{x}\right)^2 \tag{7.61}$$

which represents another circle centered at $(\Gamma_r = 1,\ \Gamma_i = 1/x)$ with a radius $1/|x|$. A family of curves (7.61) is plotted in dashed lines in Fig. 7.11. A simultaneous plot of (7.60) and (7.61) on the $\Gamma_r - \Gamma_i$ plane is called the Smith chart.

7.5 Transient Waves on Transmission Lines

Transient waves find important applications in high-frequency digital circuitry, switching transmission lines, etc. For illustration we consider a transmission line with a switch in Fig. 7.12. When the switch is closed at $t = 0$, a transient wave is generated on the transmission line for $t > 0$. Transient waves, which vary in space z and time t, can be described in terms of voltages $V(z, t)$ and currents $I(z, t)$. For notational simplicity we use V and I instead of $V(z, t)$ and $I(z, t)$ from now on. Notations V and I used in this section should not be confused with phasor notations used in other sections.

For transient wave analysis, we will devise another scheme, called a reflection diagram. Figure 7.13 illustrates a voltage reflection diagram consisting of temporal (vertical t) and spatial (horizontal z) axes. When the switch is closed, a wave starts to propagate toward the receiving end at $z = l$. Utilizing the reflection diagram, we keep track of propagating waves as follows:

1. (For $0 < t < T$) Upon closing the switch at $t = 0$, the incident wave ($V^+ = Z_0 I^+$ and I^+) propagates from the sending end at $z = 0$ with a constant speed u. To determine V^+ we replace the transmission line at $z = 0$ with an equivalent impedance Z_0, as shown in Fig. 7.14. Applying Kirchhoff's voltage law around the loop in the equivalent problem, we obtain

$$V^+ = Z_0 \underbrace{\frac{V_s}{Z_0 + Z_s}}_{I^+} . \tag{7.62}$$

The incident voltage V^+ arrives at the receiving end at $z = l$ when $t = T \, (= l/u)$, where T denotes a transit time.

2. (At $t = T$) When V^+ arrives at the load $z = l$, a reflection occurs. Let (V^+, I^+) denote the incident wave and (V^-, I^-) denote the reflected wave at $z = l$. Thus we get the total voltage and current at $z = l$ as

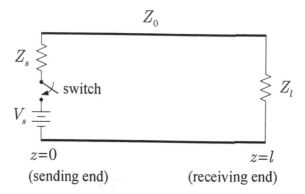

Fig. 7.12   A transmission line with a switch

Fig. 7.13 Reflection diagram
for voltage

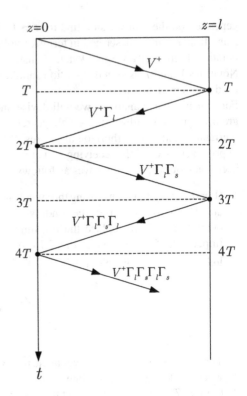

$$V = V^+ + V^- \tag{7.63}$$

$$I = \underbrace{I^+}_{\dfrac{V^+}{Z_0}} + \underbrace{I^-}_{-\dfrac{V^-}{Z_0}} . \tag{7.64}$$

For simplicity the load is assumed to be purely resistive (Z_l: real). Then we have

$$Z_l = \frac{V}{I} = Z_0 \frac{V^+ + V^-}{V^+ - V^-}. \tag{7.65}$$

The voltage reflection coefficient Γ_l in the load is

$$\Gamma_l = \frac{V^-}{V^+} = \frac{Z_l - Z_0}{Z_l + Z_0}. \tag{7.66}$$

3. (For $t > T$) The reflected voltage $V^+\Gamma_l$ propagates back from the load toward
   the sending end. At $t = 2T$, $V^+\Gamma_l$ arrives at $z = 0$ and is again reflected with
   the voltage reflection coefficient $\Gamma_s = (Z_s - Z_0)/(Z_s + Z_0)$. This reflection
   process repeats continually as time progresses. The total voltage on the trans-

Fig. 7.14 Waves at the
sending end

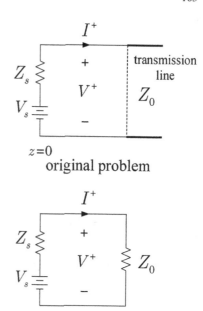

I^+

Z_s

V_s

$+$

V^+

$-$

transmission
line

Z_0

$z=0$

original problem

I^+

Z_s

V_s

$+$

V^+

$-$

Z_0

$z=0$

equivalent problem

mission line is the sum of the incident $\left[V^+,\ V^+(\Gamma_l\Gamma_s),\ \cdots\ \right]$ and reflected $\left[V^+\Gamma_l,\ V^+\Gamma_l(\Gamma_s\Gamma_l),\ \cdots\ \right]$ voltages.

4. (For $t \to \infty$) When
$t \to \infty$, the voltage is

$$V\Big|_{t \to \infty} = \underbrace{\left[V^+ + V^+(\Gamma_l\Gamma_s) + V^+(\Gamma_l\Gamma_s)^2 + \cdots \right]}_{\dfrac{V^+}{1-\Gamma_l\Gamma_s}}$$

$$+ \underbrace{\left[V^+\Gamma_l + V^+\Gamma_l(\Gamma_s\Gamma_l) + V^+\Gamma_l(\Gamma_s\Gamma_l)^2 + \cdots \right]}_{\dfrac{V^+\Gamma_l}{1-\Gamma_s\Gamma_l}}$$

$$= V_s \dfrac{Z_l}{Z_l + Z_s} \tag{7.67}$$

which shows that the source voltage V_s is divided between Z_l and Z_s. This steady-state situation is illustrated in Fig. 7.15.

Example 7.3   Voltage distributions on a line.
Sketch a voltage distribution versus time t at $z = l/2$ for a transmission line in

Fig. 7.15 Transmission line when $t \rightarrow \infty$

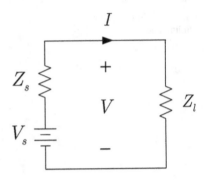

Fig. 7.12, where $Z_l = 0$ and $Z_s = 3Z_0$. Also sketch a voltage distribution versus the distance z at $t = 2.5T$, where T is a transit time.

Solution: The transient voltage and current can be obtained from the reflection diagram where

$$V^+ = V_s \frac{Z_0}{Z_s + Z_0} = \frac{V_s}{4} \tag{7.68}$$

$$\Gamma_s = \frac{Z_s - Z_0}{Z_s + Z_0} = \frac{1}{2} \tag{7.69}$$

$$\Gamma_l = \frac{Z_l - Z_0}{Z_l + Z_0} = -1. \tag{7.70}$$

The voltage reflection diagram is shown in Fig. 7.16. Utilizing the voltage reflection diagram, we obtain the total voltage versus time as

1. for $0 < t < 0.5T$, $V\big|_{z=l/2} = 0$

2. for $0.5T < t < 1.5T$, $V\big|_{z=l/2} = \dfrac{V_s}{4}$

3. for $1.5T < t < 2.5T$, $V\big|_{z=l/2} = \dfrac{V_s}{4} - \dfrac{V_s}{4} = 0$

4. for $2.5T < t < 3.5T$, $V\big|_{z=l/2} = \dfrac{V_s}{4} - \dfrac{V_s}{4} - \dfrac{V_s}{8} = -\dfrac{V_s}{8}$

5. for $3.5T < t < 4.5T$, $V\big|_{z=l/2} = \dfrac{V_s}{4} - \dfrac{V_s}{4} - \dfrac{V_s}{8} + \dfrac{V_s}{8} = 0$

6. for $4.5T < t < 5.5T$, $V\big|_{z=l/2} = \dfrac{V_s}{4} - \dfrac{V_s}{4} - \dfrac{V_s}{8} + \dfrac{V_s}{8} + \dfrac{V_s}{16} = \dfrac{V_s}{16}$.

The voltage distributions versus time and distance are given in Figs. 7.17 and 7.18, respectively.

Fig. 7.16 Reflection diagram
for voltage with $Z_l = 0$ and
$Z_s = 3Z_0$

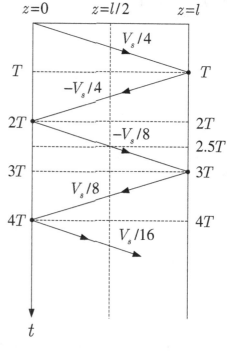

Fig. 7.17 Voltage distribution
versus time at $z = l/2$

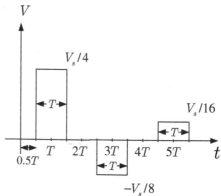

7.6 Problems for Chapter 7

1. An AC voltage source V with zero internal impedance is applied across the inner
   and outer conductors of an infinitely long coaxial cable, as shown in Fig. 7.19.
   Power flows within an annular region ($a < \rho < b$) of permittivity ϵ and permeability μ. Evaluate the time-average power P_{av} delivered by a coaxial cable.

Fig. 7.18 Voltage distribution versus the distance z at
$t = 2.5T$

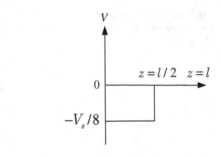

Fig. 7.19 A coaxial cable
with radii a and b

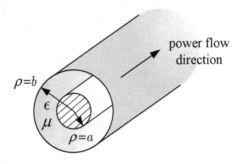

2. Consider a lossless transmission line of Fig. 7.7 in AC steady state. Obtain the
   time-average power P_{av} delivered to the load Z_l when $Z_s = Z_0$.

   Hint: $P_{av} = \dfrac{1}{2} \operatorname{Re}(V I^*)$

3. Figure 7.20 depicts a two-stage lossless transmission line of the characteristic
   impedances Z_1 and Z_2 terminated with a load impedance Z_l. The transmission
   line is in AC steady state with an AC voltage source V_s. The lengths of lines 1

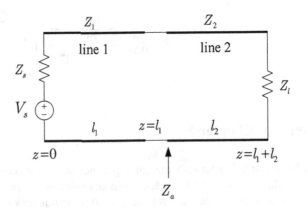

Fig. 7.20 Terminated double-section transmission line

Fig. 7.21 A rectangular pulse
shape

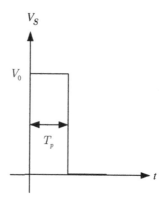

Fig. 7.22 Transmission line
of characteristic impedance
Z_0 with pulse excitation

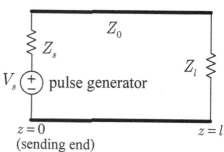

and 2 are $l_1 (= \lambda_1)$ and $l_2 (= \lambda_2/2)$, where λ_1 and λ_2 are the wavelengths on lines 1 and 2, respectively. Evaluate the voltage reflection coefficient at $z = l_1/2$.
Hint: Evaluate the impedance Z_a by starting from the load at $z = l_1 + l_2$.

4. Evaluate the input impedance of a shorted transmission line of quarter-wave length $(\lambda/4)$ by using the Smith chart.

5. Figure 7.21 shows a rectangular voltage pulse V_s with a duration T_p. This pulse is transmitted from the sending end at $t = 0$, as illustrated in Fig. 7.22. Sketch a voltage distribution versus t at $z = l/2$ when $Z_s = Z_0 = Z_l$.
Hint: Convert the pulse into a linear combination of two unit step functions and construct a reflection diagram.

Chapter 8
Waveguides and Antennas

8.1 Fundamentals of Waveguides

Waveguides and transmission lines are widely used to guide electromagnetic waves
from one point to another. Unlike transmission lines guiding TEM waves with two
conducting terminals, waveguides can guide higher-order modes. Two representative
waveguide examples are rectangular waveguides and optical fibers that are used in
radio wave and optical communications. Figure 8.1 shows a cylindrical waveguide
of constant cross section guiding electromagnetic waves along the z-direction. Then
guided waves obey the source-free Maxwell's equations

$$\nabla \times \overline{E} = -j\omega\mu\overline{H} \tag{8.1}$$

$$\nabla \times \overline{H} = j\omega\epsilon\overline{E}. \tag{8.2}$$

Our aim is to determine guided waves by solving Maxwell's equations subject to
prescribed boundary conditions. When waves are guided along the z-direction, they
exhibit a z-dependance $\exp(-jk_z z)$ and their field components are written as

$$\underbrace{E(x, y, z)}_{E} = \underbrace{\tilde{E}(x, y)}_{\tilde{E}} \exp(-jk_z z). \tag{8.3}$$

It is possible to rewrite (8.1) and (8.2) in six scalar equations using rectangular
coordinates as follows:

H. J. Eom, *Primary Theory of Electromagnetics*, Power Systems,
DOI: 10.1007/978-94-007-7143-7_8, © Springer Science+Business Media Dordrecht 2013

Fig. 8.1 Waveguide of a
uniform cross section guiding
fields $\overline{E}$ and $\overline{H}$ along the
z-direction

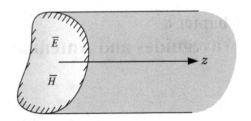

$$\frac{\partial E_z}{\partial y} + jk_z E_y = -j\omega\mu H_x \tag{8.4}$$

$$-jk_z E_x - \frac{\partial E_z}{\partial x} = -j\omega\mu H_y \tag{8.5}$$

$$\frac{\partial E_y}{\partial x} - \frac{\partial E_x}{\partial y} = -j\omega\mu H_z \tag{8.6}$$

$$\frac{\partial H_z}{\partial y} + jk_z H_y = j\omega\epsilon E_x \tag{8.7}$$

$$-jk_z H_x - \frac{\partial H_z}{\partial x} = j\omega\epsilon E_y \tag{8.8}$$

$$\frac{\partial H_y}{\partial x} - \frac{\partial H_x}{\partial y} = j\omega\epsilon E_z. \tag{8.9}$$

The transverse components H_x, H_y, E_x, and E_y can be written in terms of the longitudinal components H_z and E_z. For example, the representation of H_x is obtained by manipulating (8.4) and (8.8). We summarize all these transverse field components as

$$H_x = \frac{j}{(k^2 - k_z^2)}\left(-k_z\frac{\partial H_z}{\partial x} + \omega\epsilon\frac{\partial E_z}{\partial y}\right) \tag{8.10}$$

$$H_y = \frac{-j}{(k^2 - k_z^2)}\left(k_z\frac{\partial H_z}{\partial y} + \omega\epsilon\frac{\partial E_z}{\partial x}\right) \tag{8.11}$$

$$E_x = \frac{-j}{(k^2 - k_z^2)}\left(k_z\frac{\partial E_z}{\partial x} + \omega\mu\frac{\partial H_z}{\partial y}\right) \tag{8.12}$$

$$E_y = \frac{j}{(k^2 - k_z^2)}\left(-k_z\frac{\partial E_z}{\partial y} + \omega\mu\frac{\partial H_z}{\partial x}\right). \tag{8.13}$$

transverse fields in terms of H_z and E_z

Here k $(= \omega\sqrt{\mu\epsilon})$ is the wave number, where μ and ϵ are the permeability and permittivity of medium inside the waveguide. We will assume that μ and ϵ are constant. The longitudinal fields E_z and H_z satisfy Helmholtz's equations

$$\frac{\partial^2 E_z}{\partial x^2} + \frac{\partial^2 E_z}{\partial y^2} + \underbrace{\frac{\partial^2 E_z}{\partial z^2} + k^2 E_z}_{(k^2 - k_z^2)E_z} = 0 \tag{8.14}$$

$$\frac{\partial^2 H_z}{\partial x^2} + \frac{\partial^2 H_z}{\partial y^2} + \underbrace{\frac{\partial^2 H_z}{\partial z^2} + k^2 H_z}_{(k^2 - k_z^2)H_z} = 0. \tag{8.15}$$

Once E_z and H_z are determined from (8.14) and (8.15), the evaluation of transverse field components is straightforward. Depending on the types of boundary conditions imposed on the waveguide boundaries, waveguides can admit different wave types. Some of them are listed below:

1. When $E_z = 0$ and $H_z \neq 0$, this wave is called a TE wave (transverse electric to the propagation direction z). The TE wave implies that the electric field vector is transverse to the propagation direction.
2. When $H_z = 0$ and $E_z \neq 0$, this wave is called a TM wave (transverse magnetic to the propagation direction z). The TM wave implies that the magnetic field vector is transverse to the propagation direction. Metallic waveguides of a uniform cross section can support both TM and TE waves.
3. When $E_z = H_z = 0$, the transverse field components (H_x, H_y, E_x, and E_y) need rewriting. From (8.4), (8.5), (8.7), and (8.8), we get the relations $E_x/H_y = \sqrt{\mu/\epsilon}$, $E_y/H_x = -\sqrt{\mu/\epsilon}$, and $k_z = k$. This wave is called a TEM wave (transverse electromagnetic to the propagation direction z). The TEM wave implies that the electric field vector and the magnetic field vector are both transverse to the propagation direction. TEM waves are guided by transmission lines that consist of two conducting terminals such as coaxial cables and paired wires.

8.2 Rectangular Waveguides

Metallic rectangular waveguides are basic components that are used to transmit microwaves and millimeter waves in various high-frequency applications. Metallic rectangular waveguides can guide either TM waves or TE waves. TM or TE waves can be determined by solving Helmholtz's equations subject to metallic boundary conditions. TM waves will be investigated first.

8.2.1 TM Waves

Consider TM wave propagation within a metallic rectangular waveguide in Fig. 8.2. A TM wave is characterized by the field configuration $\left(H_z = 0 \text{ and } E_z = \tilde{E}_z e^{-jk_z z}\right)$,

Fig. 8.2 Metallic rectangular
waveguide with dimensions
$a \times b$

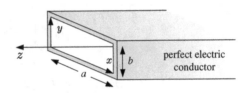

perfect electric
conductor

where $\tilde{E}_z$ obeys

$$\frac{\partial^2 \tilde{E}_z}{\partial x^2} + \frac{\partial^2 \tilde{E}_z}{\partial y^2} + (k^2 - k_z^2)\tilde{E}_z = 0. \tag{8.16}$$

To find $\tilde{E}_z$, we will use the method of separation of variables, which assumes a product form $\tilde{E}_z = XY$, where X and Y are functions of x and y, respectively. Substituting XY into (8.16) and dividing (8.16) by XY, we obtain

$$\underbrace{\frac{1}{X}\frac{d^2X}{dx^2}}_{-k_x^2} + \underbrace{\frac{1}{Y}\frac{d^2Y}{dy^2}}_{-k_y^2} + k^2 - k_z^2 = 0. \tag{8.17}$$

To make (8.17) valid for every x and y, the separation parameters k_x and k_y should be constants and

$$k_x^2 + k_y^2 + k_z^2 = k^2. \tag{8.18}$$

We have

$$\frac{1}{X}\frac{d^2X}{dx^2} + k_x^2 = 0 \Longrightarrow X = c_1 \cos k_x x + c_2 \sin k_x x \tag{8.19}$$

$$\frac{1}{Y}\frac{d^2Y}{dy^2} + k_y^2 = 0 \Longrightarrow Y = c_3 \cos k_y y + c_4 \sin k_y y. \tag{8.20}$$

Hence the solution is written as

$$E_z = (c_1 \cos k_x x + c_2 \sin k_x x)(c_3 \cos k_y y + c_4 \sin k_y y) e^{-jk_z z}. \tag{8.21}$$

To determine the field, the boundary conditions on the metallic surfaces must be used. The boundary conditions require the electric fields tangential to the conducting walls to vanish as

1. The condition $E_z|_{x=0} = E_y|_{x=0} = 0$ gives $c_1 = 0$.
2. The condition $E_z|_{x=a} = E_y|_{x=a} = 0$ gives $k_x = \dfrac{m\pi}{a}$, where $m = 1, 2, 3, \cdots$.
3. Similarly the condition $E_z|_{y=0} = E_x|_{y=0} = 0$ gives $c_3 = 0$.
4. The condition $E_z|_{y=b} = E_x|_{y=b} = 0$ gives $k_y = \dfrac{n\pi}{b}$, where $n = 1, 2, 3, \cdots$.

Therefore the E_z component is written as

$$E_z = E_0 \sin\left(\frac{m\pi}{a}x\right) \sin\left(\frac{n\pi}{b}y\right) e^{-jk_z z} \tag{8.22}$$

where E_0 is a field amplitude. The rest of transverse field components are

$$
\begin{aligned}
H_x &= \frac{j}{(k^2 - k_z^2)}\left(\omega\epsilon\frac{\partial E_z}{\partial y}\right)\\
&= \frac{j\omega\epsilon\left(\frac{n\pi}{b}\right)}{\left(\frac{m\pi}{a}\right)^2 + \left(\frac{n\pi}{b}\right)^2} E_0 \sin\left(\frac{m\pi}{a}x\right)\cos\left(\frac{n\pi}{b}y\right) e^{-jk_z z}
\end{aligned} \tag{8.23}
$$

$$
\begin{aligned}
H_y &= \frac{-j}{(k^2 - k_z^2)}\left(\omega\epsilon\frac{\partial E_z}{\partial x}\right)\\
&= \frac{-j\omega\epsilon\left(\frac{m\pi}{a}\right)}{\left(\frac{m\pi}{a}\right)^2 + \left(\frac{n\pi}{b}\right)^2} E_0 \cos\left(\frac{m\pi}{a}x\right)\sin\left(\frac{n\pi}{b}y\right) e^{-jk_z z}
\end{aligned} \tag{8.24}
$$

$$
\begin{aligned}
E_x &= \frac{-j}{(k^2 - k_z^2)}\left(k_z\frac{\partial E_z}{\partial x}\right)\\
&= \frac{-jk_z\left(\frac{m\pi}{a}\right)}{\left(\frac{m\pi}{a}\right)^2 + \left(\frac{n\pi}{b}\right)^2} E_0 \cos\left(\frac{m\pi}{a}x\right)\sin\left(\frac{n\pi}{b}y\right) e^{-jk_z z}
\end{aligned} \tag{8.25}
$$

$$
\begin{aligned}
E_y &= \frac{j}{(k^2 - k_z^2)}\left(-k_z\frac{\partial E_z}{\partial y}\right)\\
&= \frac{-jk_z\left(\frac{n\pi}{b}\right)}{\left(\frac{m\pi}{a}\right)^2 + \left(\frac{n\pi}{b}\right)^2} E_0 \sin\left(\frac{m\pi}{a}x\right)\cos\left(\frac{n\pi}{b}y\right) e^{-jk_z z}.
\end{aligned} \tag{8.26}
$$

The field described by the above expressions is referred to as a TM$_{mn}$ mode. Any linear combination of TM$_{mn}$ modes can be a valid representation of waves propagating in rectangular waveguides. Wave characteristics along the z-direction are governed by k_z. Care must be exercised in evaluating k_z since the wave should be properly propagating or evanescent along the z-direction. To this end, we let

$$k_z = \begin{cases} \sqrt{k^2 - k_{cmn}^2} & \text{for } k > k_{cmn} \\ -j\sqrt{k_{cmn}^2 - k^2} & \text{for } k < k_{cmn} \end{cases} \tag{8.27}$$

where $k_{cmn} = \sqrt{(m\pi/a)^2 + (n\pi/b)^2}$ is called a cutoff wave number. The choice of k_z for $k < k_{cmn}$ makes the wave $e^{-jk_z z}$ evanescent when $z \to \infty$. We introduce a cutoff frequency f_{cmn} for the TM$_{mn}$ mode as

$$\begin{aligned} f_{cmn} &= \frac{k_{cmn}}{2\pi\sqrt{\mu\epsilon}} \\ &= \frac{1}{2\sqrt{\mu\epsilon}}\sqrt{(m/a)^2 + (n/b)^2}. \end{aligned} \tag{8.28}$$

The TM$_{mn}$ mode propagates when the operating frequency is higher than f_{cmn}, but it becomes evanescent when the frequency is lower than f_{cmn}. The lowest-order TM wave is the TM$_{11}$ mode.

8.2.2 TE Waves

Metallic rectangular waveguides can also guide TE waves $(E_z = 0$ and $H_z \neq 0)$. A TE wave analysis is somewhat similar to the TM wave case. The magnetic field H_z satisfies Helmholtz's equation

$$\frac{\partial^2 H_z}{\partial x^2} + \frac{\partial^2 H_z}{\partial y^2} + \frac{\partial^2 H_z}{\partial z^2} + k^2 H_z = 0 \tag{8.29}$$

subject to the boundary conditions requiring zero tangential electric fields on metallic surfaces. The boundary conditions are given by

$$E_x\big|_{y=0,b} = -\frac{j\omega\mu}{k^2 - k_z^2}\frac{\partial H_z}{\partial y}\bigg|_{y=0,b} = 0 \tag{8.30}$$

$$E_y\big|_{x=0,a} = \frac{j\omega\mu}{k^2 - k_z^2}\frac{\partial H_z}{\partial x}\bigg|_{x=0,a} = 0. \tag{8.31}$$

Based on the method of separation of variables, H_z takes the form of

$$H_z = H_0 \cos k_x x \, \cos k_y y \, \exp(-jk_z z) \tag{8.32}$$

where H_0 is a field amplitude, $k_x = m\pi/a$, and $k_y = n\pi/b$. The parameters m and n are integers ($m, n = 0, 1, 2, 3, \cdots$: m and n cannot be zero simultaneously). Note that k_z is given by (8.27). Fields of TE$_{mn}$ mode are obtained from (8.32). The TE$_{mn}$

mode propagates when $k > k_{cmn}$, but it becomes evanescent when $k < k_{cmn}$. The lowest-order wave propagating within a rectangular waveguide, when $a > b$, is the TE_{10} mode.

Example 8.1 TE_{10} **mode.**

Evaluate the time-average power of a TE_{10} mode propagating in a rectangular waveguide with dimensions $a > b$ when $(\pi/a) < k$.

Solution: The nonzero-field components are

$$H_z = H_0 \cos k_x x \, e^{-jk_z z} \tag{8.33}$$

$$H_x = \frac{jk_z}{k_x} H_0 \sin k_x x \, e^{-jk_z z} \tag{8.34}$$

$$E_y = -\frac{j\omega\mu}{k_x} H_0 \sin k_x x \, e^{-jk_z z} \tag{8.35}$$

where $k_x = \pi/a$ and $k_z = \sqrt{k^2 - (\pi/a)^2}$. The time-average power density delivered by a TE_{10} mode is

$$\overline{S}_{av} = \frac{1}{2} \operatorname{Re}\left(\overline{E} \times \overline{H}^*\right) = \frac{1}{2} \operatorname{Re}\left(-\hat{z} E_y H_x^* + \hat{x} E_y H_z^*\right). \tag{8.36}$$

The time-average power delivery along the z-direction is obtained by integrating $\overline{S}_{av}$ over a waveguide cross section $(a \times b)$:

$$
\begin{aligned}
P_{av} &= \int_0^b \int_0^a \overline{S}_{av} \cdot \hat{z} \, dx \, dy \\
&= -\int_0^b \int_0^a \frac{1}{2} \operatorname{Re}\left(E_y H_x^*\right) dx \, dy \\
&= \frac{\omega\mu k_z a^3 b}{4\pi^2} |H_0|^2.
\end{aligned}
\tag{8.37}
$$

8.3 Fundamentals of Antennas

Antennas transmit and receive electromagnetic waves. Electromagnetic wave radiation from time-varying currents is a basic mechanism of antennas. Transmitting antennas convert currents into electromagnetic waves and receiving antennas conversely convert electromagnetic waves into currents. In this section, we will present two important fundamental concepts for antenna analysis: free-space solution and far fields.

Fig. 8.3 Current density $\overline{J}'$
in free space

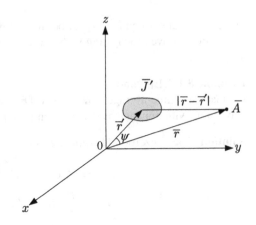

8.3.1 Free-Space Solution

Simple antennas are usually modeled as currents that radiate electromagnetic waves.
Consider a current density $\overline{J}'\ (=\overline{J}(\overline{r}'))$ in free space, as illustrated in Fig. 8.3. Here $\overline{r}'$
denotes a position vector designating the current source distribution. Electromagnetic
waves and the current density are related by Helmholtz's equation for the magnetic
vector potential $\overline{A}$ as

$$\nabla^2\overline{A} + k^2\overline{A} = -\mu\overline{J} \tag{8.38}$$

where $\overline{A}$ and $\overline{J}$ are functions of $\overline{r}$ such as $\overline{A} = \overline{A}(\overline{r})$ and $\overline{J} = \overline{J}(\overline{r})$. The solution $\overline{A}$,
when $\overline{J}$ is in free space, is usually called a free-space solution. We will derive the
free-space solution by using a Green's function method.

Green's function: To solve (8.38) for $\overline{A}$, we first introduce a free-space Green's
function G that satisfies

$$\nabla^2 G + k^2 G = -\delta\left(\overline{r} - \overline{r}'\right) \tag{8.39}$$

where $\delta\left(\overline{r} - \overline{r}'\right)$ denotes the three-dimensional Dirac delta function. For instance in
rectangular coordinates, this is given by

$$\delta\left(\overline{r} - \overline{r}'\right) = \delta(x - x')\,\delta(y - y')\,\delta(z - z'). \tag{8.40}$$

The Green's function G represents a response at $\overline{r}$ due to the delta source at $\overline{r}'$, as
shown in Fig. 8.4. Let us introduce the spherical coordinates (R, ϑ, φ) centered at $\overline{r}'$
where

$$x - x' = R\sin\vartheta\cos\varphi \tag{8.41}$$
$$y - y' = R\sin\vartheta\sin\varphi \tag{8.42}$$
$$z - z' = R\cos\vartheta \tag{8.43}$$

Fig. 8.4 A delta-source
response G in free space

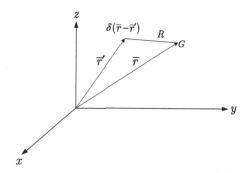

and $R = |\bar{r} - \bar{r}'|$. Then due to symmetry with respect to the origin $R = 0$, G becomes a function of R only, and (8.39) for $R \neq 0$ is rewritten as

$$\underbrace{\frac{1}{R^2}\frac{d}{dR}\left(R^2\frac{dG}{dR}\right)}_{\nabla^2 G} + k^2 G = 0. \tag{8.44}$$

With an educated guess we substitute $G = \dfrac{U}{R}$ into (8.44) to obtain

$$\frac{d^2U}{dR^2} + k^2 U = 0 \tag{8.45}$$

whose solution is

$$U = c_1 e^{-jkR} + c_2 e^{jkR}. \tag{8.46}$$

Since U must be a wave that propagates outwardly from $R = 0$, we choose

$$U = c_1 e^{-jkR} \implies G = c_1 \frac{e^{-jkR}}{R}. \tag{8.47}$$

To determine the unknown coefficient c_1, a boundary condition is usually imposed near $R = 0$. Instead of imposing the boundary condition, we will adopt a simple alternative, which is based on Poisson's equation. Note that (8.47) is a general solution that satisfies Helmholtz's equation (8.39) for every k. In particular, if $k \to 0$, (8.39) is reduced to Poisson's equation

$$\nabla^2 G = -\delta\left(\bar{r} - \bar{r}'\right) \implies G = \frac{c_1}{R}. \tag{8.48}$$

Recall an electrostatic problem of a point charge in free space. When a static point charge q is placed at $\bar{r}'$, a scalar potential V at $\bar{r}$ is given by Poisson's equation

$$\nabla^2 V = -\frac{q}{\epsilon_0} \delta\left(\bar{r} - \bar{r}'\right) \Longrightarrow V = \frac{q}{4\pi\epsilon_0 R}. \tag{8.49}$$

Upon comparing (8.48) with (8.49), we obtain $c_1 = \frac{1}{4\pi}$. Hence the free-space Green's function is

$$G = \frac{e^{-jk|\bar{r}-\bar{r}'|}}{4\pi|\bar{r} - \bar{r}'|}. \tag{8.50}$$

In order to obtain $\bar{A}$ from G, we will use the superposition. To this end we rewrite $\bar{J}(\bar{r})$ as

$$\bar{J}(\bar{r}) = \int_{v'} \bar{J}(\bar{r}')\delta(\bar{r} - \bar{r}')\,dv' \tag{8.51}$$

which is a sum of the delta sources $\bar{J}(\bar{r}')\delta(\bar{r} - \bar{r}')$ that are continuously distributed over v'. Based on the superposition, the solution to (8.38) is given by

$$\bar{A}(\bar{r}) = \mu \int_{v'} \bar{J}(\bar{r}') \frac{e^{-jk|\bar{r}-\bar{r}'|}}{4\pi|\bar{r} - \bar{r}'|}\,dv'. \tag{8.52}$$

free-space solution to Helmholtz's equation

This is a formal solution of the magnetic vector potential $\bar{A}(\bar{r})$, which results from the current density $\bar{J}(\bar{r}')$ occupying a volume v' in free space. In a static limit ($k \to 0$), $\bar{A}(\bar{r})$ is reduced to

$$\bar{A}(\bar{r}) = \mu \int_{v'} \frac{\bar{J}(\bar{r}')}{4\pi|\bar{r} - \bar{r}'|}\,dv' \tag{8.53}$$

where $\bar{A}(\bar{r})$ and $\bar{J}(\bar{r}')$ are all static quantities. Note that (8.53) is a well-known solution to Poisson's equation

$$\nabla^2 \bar{A} = -\mu\bar{J}. \tag{8.54}$$

8.3.2 Far Fields

The free-space solution to Helmholtz's equation allows us to calculate radiation fields when antennas radiate in unbounded media. We will discuss how radiated fields behave in lossless media (μ, ϵ: real numbers) when the radiated fields are reasonably far away from transmitting antennas. Consider the antenna geometry in Fig. 8.3, where the current density $\bar{J}'$ at $\bar{r}'$ represents a transmitting antenna. It is of practical interest to study antenna radiation fields at $\bar{r}$ in the far zone, which is characterized by

$$kr \gg 1 \tag{8.55}$$
$$r \gg r'. \tag{8.56}$$

far-zone condition

Here r designates the distance from the coordinate origin to the observation point, while r' designates the distance from the coordinate origin to the antenna. The position vectors $\bar{r}$ and $\bar{r}'$ are represented in terms of respective spherical coordinates as

$$\bar{r} = \hat{x} r \sin \theta \cos \phi + \hat{y} r \sin \theta \sin \phi + \hat{z} r \cos \theta \tag{8.57}$$
$$\bar{r}' = \hat{x} r' \sin \theta' \cos \phi' + \hat{y} r' \sin \theta' \sin \phi' + \hat{z} r' \cos \theta'. \tag{8.58}$$

In the far zone we obtain

$$|\bar{r} - \bar{r}'| = \sqrt{r^2 + r'^2 - 2rr' \cos \psi} \approx r - r' \cos \psi \tag{8.59}$$

where ψ denotes the angle between $\bar{r}$ and $\bar{r}'$ and

$$\cos \psi = \cos \theta \cos \theta' + \sin \theta \sin \theta' \cos(\phi - \phi'). \tag{8.60}$$

Therefore we have

$$\frac{e^{-jk|\bar{r} - \bar{r}'|}}{|\bar{r} - \bar{r}'|} \approx \frac{1}{r} e^{-jkr + jkr' \cos \psi}. \tag{8.61}$$

The magnetic vector potential in the far zone is written as

$$\bar{A} \approx \frac{\mu e^{-jkr}}{4\pi r} \underbrace{\int_{v'} \bar{J}' e^{jkr' \cos \psi} \, dv'}_{\hat{r} A'_r + \hat{\theta} A'_\theta + \hat{\phi} A'_\phi}$$
$$= \hat{r} A_r + \hat{\theta} A_\theta + \hat{\phi} A_\phi \tag{8.62}$$

where A'_r, A'_θ, and A'_ϕ are functions of θ and ϕ. The electric field is given by the magnetic vector potential as

$$\bar{E} = -j\omega \bar{A} - \frac{j}{\omega \mu \epsilon} \nabla (\nabla \cdot \bar{A}). \tag{8.63}$$

In the far zone,

$$\frac{j}{\omega\mu\epsilon}\nabla\left(\nabla\cdot\overline{A}\right)\approx\hat{r}\underbrace{\frac{j}{\omega\mu\epsilon}\frac{\partial}{\partial r}\left[\frac{1}{r^2}\frac{\partial}{\partial r}(r^2 A_r)\right]}_{-\omega^2\mu\epsilon\frac{\mu e^{-jkr}}{4\pi r}A'_r}$$

$$=-\hat{r}\,j\omega A_r. \tag{8.64}$$

The field in the far zone is called the far field. The far field $\overline{E}$ is

$$\overline{E}\approx\hat{\theta}\underbrace{(-j\omega A_\theta)}_{E_\theta}+\hat{\phi}\underbrace{(-j\omega A_\phi)}_{E_\phi}. \tag{8.65}$$

electric field in the far zone

The magnetic field is

$$\overline{H}=\frac{1}{\mu}\nabla\times\overline{A}$$

$$=\frac{1}{\mu}\left\{\frac{\hat{r}}{r\sin\theta}\left[\frac{\partial(A_\phi\sin\theta)}{\partial\theta}-\frac{\partial A_\theta}{\partial\phi}\right]\right.$$

$$\left.+\frac{\hat{\theta}}{r}\left[\frac{1}{\sin\theta}\frac{\partial A_r}{\partial\phi}-\frac{\partial(r A_\phi)}{\partial r}\right]+\frac{\hat{\phi}}{r}\left[\frac{\partial(r A_\theta)}{\partial r}-\frac{\partial A_r}{\partial\theta}\right]\right\}. \tag{8.66}$$

Since

$$\frac{\partial(r A_\phi)}{\partial r}\approx\frac{\partial(e^{-jkr})}{\partial r}\frac{\mu}{4\pi}A'_\phi=-jkr A_\phi \tag{8.67}$$

$$\frac{\partial(r A_\theta)}{\partial r}\approx-jkr A_\theta \tag{8.68}$$

(8.66) becomes

$$\overline{H}\approx\underbrace{\frac{jk}{\mu}\left(\hat{\theta}A_\phi-\hat{\phi}A_\theta\right)}_{1/r-\text{decaying terms}}+\left(1/r^2-\text{decaying terms}\right). \tag{8.69}$$

Ignoring the $1/r^2$-decaying terms in the far zone ($kr\gg 1$), we obtain

$$\overline{H} \approx \hat{\theta}\,(\underbrace{\frac{jk}{\mu}A_\phi}_{H_\theta}) + \hat{\phi}\,(\underbrace{-\frac{jk}{\mu}A_\theta}_{H_\phi})\,. \tag{8.70}$$

magnetic field in the far zone

Note that $\overline{E}\cdot\overline{H} = 0$, indicating that the far fields $\overline{E}$ and $\overline{H}$ are mutually perpendicular. The far fields satisfy the relation

$$\overline{H} = \frac{\hat{r}}{\eta} \times \overline{E} \tag{8.71}$$

TEM wave in the far zone

where $\hat{r}$ is a unit vector in the $\overline{r}$-direction and $\eta\,(=\sqrt{\mu/\epsilon})$ is the intrinsic impedance of a medium, which is $120\,\pi\,(\Omega)$ in air. Far fields are shown to be TEM waves with the θ- and ϕ-components propagating in the $\overline{r}$-direction. Utilizing the time-average Poynting vector $\frac{1}{2}\,\mathrm{Re}\,\left(\overline{E}\times\overline{H}^*\right)$, we evaluate the total time-average power radiated through a spherical surface s:

$$\begin{aligned} P_{rad} &= \oint_s \frac{1}{2}\,\mathrm{Re}\,\left(\overline{E}\times\overline{H}^*\right)\cdot d\overline{s} \\ &= \int_0^{2\pi}\int_0^{\pi}\Big[\underbrace{\frac{1}{2\eta}\left(|E_\theta|^2 + |E_\phi|^2\right)r^2}_{U(\theta,\,\phi)}\Big]\sin\theta\,d\theta\,d\phi \end{aligned} \tag{8.72}$$

antenna radiation power

where $U(\theta,\,\phi)$ is called the radiation intensity. Since the far field (E_θ and E_ϕ) decays as $1/r$, the radiation intensity $U(\theta,\,\phi)$ becomes a function of $(\theta,\,\phi)$ independent of r. The radiation intensity is the time-average power radiated per unit solid angle (differential solid angle $= \sin\theta\,d\theta\,d\phi$). Hence, the average of $U(\theta,\,\phi)$ is given as

$$U_{av} = \frac{P_{rad}}{4\pi}\,. \tag{8.73}$$

When the antenna radiates $U(\theta,\,\phi)$ in a direction $(\theta,\,\phi)$, then the ratio of $U(\theta,\,\phi)$ to U_{av} is called the directivity of the antenna

Fig. 8.5 Radiation from a
Hertzian dipole

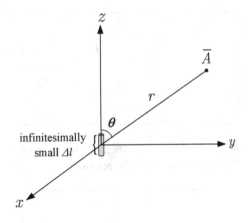

$$D = \frac{U(\theta, \phi)}{U_{av}} = \frac{4\pi U(\theta, \phi)}{P_{rad}}. \tag{8.74}$$

The directivity is a measure of antenna ability to radiate in a given direction (θ, ϕ).

8.4 Wire Antennas

Real, practical antennas cover diverse types from simple to complex geometrical structures. Complex geometrical structures are often required to improve antenna performances in gain, bandwidth, or radiation efficiency. Simple antenna types are thin wires such as dipole, monopole, and loop types. These thin wire antennas can be modeled in terms of line currents. In this section, we will discuss the radiation characteristics for Hertzian dipole and small circular loop antennas.

8.4.1 Hertzian Dipoles

Figure 8.5 illustrates an infinitesimal current density given by

$$\overline{J}' = \hat{z} I \, \Delta l \, \delta(x') \, \delta(y') \, \delta(z') \tag{8.75}$$

where I is a current, Δl is a small length compared with the wavelength, and $\delta(\cdot)$ is the Dirac delta function. The infinitesimal current density of delta functions given by (8.75) is known as a Hertzian dipole. Since any arbitrary currents can be decomposed into a sum of Hertzian dipoles, a Hertzian-dipole concept is useful for antenna modeling. The magnetic vector potential of a Hertzian dipole in the far zone is

Fig. 8.6 Radiation from a
circular loop antenna

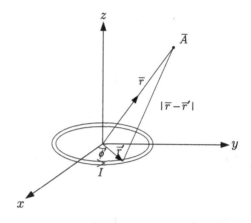

$$\overline{A} = \frac{\mu e^{-jkr}}{4\pi r} \underbrace{\int_{v'} \overline{J}' e^{jkr' \cos\psi} \, dv'}_{\hat{z} I \Delta l}$$

$$= \frac{\mu e^{-jkr}}{4\pi r} I \Delta l \underbrace{(\hat{r} \cos\theta - \hat{\theta} \sin\theta)}_{\hat{z}} . \tag{8.76}$$

The far fields are

$$E_\theta = -j\omega A_\theta = j\omega \frac{\mu e^{-jkr}}{4\pi r} I \Delta l \sin\theta \tag{8.77}$$

$$H_\phi = \frac{E_\theta}{\eta} \tag{8.78}$$

where $\eta = \sqrt{\mu/\epsilon}$ is the medium intrinsic impedance. Maximum radiation occurs at $\theta = 90°$ and the total radiated power is

$$P_{rad} = \int_0^{2\pi} d\phi \int_0^{\pi} \frac{1}{2\eta} |E_\theta|^2 r^2 \sin\theta \, d\theta = \frac{\eta}{12\pi} (k\Delta l)^2 |I|^2. \tag{8.79}$$

8.4.2 Circular Loop Antennas

Figure 8.6 illustrates an infinitesimally thin circular loop antenna of radius a carrying a constant current I. Assume that the current density for a small loop antenna is

$$\overline{J}' = \hat{\phi}' I \, \delta(\rho' - a) \, \delta(z') \tag{8.80}$$

where $\delta(\cdot)$ denotes the Dirac delta function. We write the magnetic vector potential of the loop current as

$$
\begin{aligned}
\bar{A} &= \frac{\mu e^{-jkr}}{4\pi r} \int_{v'} \bar{J}' \, e^{jkr' \cos\psi} \, dv' \\
&= \frac{\mu e^{-jkr}}{4\pi r} I \int_0^{2\pi} \hat{\phi}' \, e^{jkr' \cos\psi} a \, d\phi'
\end{aligned}
\tag{8.81}
$$

where ψ denotes the angle between $\bar{r}$ and $\bar{r}'$ and

$$
r' \cos\psi = a \sin\theta \cos(\phi - \phi').
\tag{8.82}
$$

Care must be exercised in handling $\hat{\phi}'$ since the direction $\hat{\phi}'$ varies as ϕ' varies. Since the directions of $\hat{x}$ and $\hat{y}$ are invariant, it is convenient to convert $\hat{\phi}'$ into $\hat{x}$ and $\hat{y}$ as

$$
\hat{\phi}' = -\hat{x} \sin\phi' + \hat{y} \cos\phi'.
\tag{8.83}
$$

We evaluate the components of $\bar{A} \ (= \hat{x} A_x + \hat{y} A_y)$ as

$$
A_x = -\frac{\mu e^{-jkr}}{4\pi r} I \int_0^{2\pi} \sin\phi' \exp\left[jka \sin\theta \cos(\phi - \phi')\right] a \, d\phi'
\tag{8.84}
$$

$$
A_y = \frac{\mu e^{-jkr}}{4\pi r} I \int_0^{2\pi} \cos\phi' \exp\left[jka \sin\theta \cos(\phi - \phi')\right] a \, d\phi'.
\tag{8.85}
$$

If the antenna size is much smaller than the wavelength ($ka \ll 1$), then

$$
\exp\left[jka \sin\theta \cos(\phi - \phi')\right] \approx 1 + jka \sin\theta \cos(\phi - \phi').
\tag{8.86}
$$

Hence

$$
A_x = -j\frac{ka^2 \mu I e^{-jkr}}{4r} \sin\theta \sin\phi
\tag{8.87}
$$

$$
A_y = j\frac{ka^2 \mu I e^{-jkr}}{4r} \sin\theta \cos\phi.
\tag{8.88}
$$

Combining A_x and A_y, we obtain

$$
\bar{A} = j\frac{ka^2 \mu I e^{-jkr}}{4r} \sin\theta \underbrace{(-\hat{x} \sin\phi + \hat{y} \cos\phi)}_{\hat{\phi}}.
\tag{8.89}
$$

Therefore the far field is

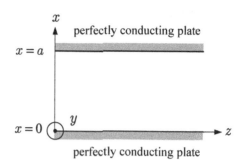

Fig. 8.7 Conducting parallel plates

$$E_\phi = -j\omega A_\phi = \frac{e^{-jkr}}{4r}(ka)^2 I\eta \sin\theta \qquad (8.90)$$

where $\eta = \sqrt{\mu/\epsilon}$ is the medium intrinsic impedance. Maximum antenna radiation takes place at $\theta = 90°$ and the total radiated power is given by

$$P_{rad} = \int_0^{2\pi} d\phi \int_0^\pi \frac{1}{2\eta}|E_\phi|^2 r^2 \sin\theta \, d\theta$$

$$= \frac{\pi}{12}\eta(ka)^4|I|^2. \qquad (8.91)$$

8.5 Problems for Chapter 8

1. Consider an air-filled rectangular waveguide with the dimensions (2.3 cm × 1.0 cm). Determine the range of operating frequencies that guarantee single-mode propagation.
2. Evaluate the time-average power carried by a TM_{mn} mode propagating within a rectangular waveguide of cross-sectional dimensions $(a \times b)$.
   Hint: The field is given by $E_z = E_0 \sin\left(\frac{m\pi}{a}x\right) \sin\left(\frac{n\pi}{b}y\right)e^{-jk_z z}$. Use the Poynting vector.
3. Figure 8.7 shows perfectly conducting parallel plates guiding TM waves ($H_z = 0$, $E_z \neq 0$) in the z-direction. Assuming the waves are independent of y, determine the TM_m mode components.
   Hint: The TM wave $\left(E_z = \tilde{E}_z e^{-jk_z z}\right)$ satisfies Helmholtz's equation $\frac{d^2\tilde{E}_z}{dx^2} + (k^2 - k_z^2)\tilde{E}_z = 0$.
4. The original problem in Fig. 8.8 illustrates a Hertzian dipole

$$\bar{J} = \hat{z}I \, \Delta l \, \delta(x) \, \delta(y) \, \delta(z - h) \qquad (8.92)$$

Fig. 8.8 A Hertzian dipole
above a perfectly conducting
plane

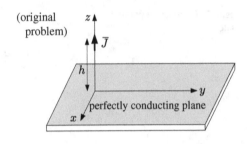

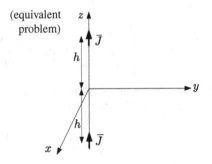

Fig. 8.9 Hertzian dipole array

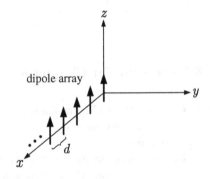

located above a perfectly conducting plane of infinite extent. Determine the far
field at $z > 0$.

Hint: Convert the original problem into its equivalent by using the image method.
The magnetic vector potential A_z is a sum of the real and image currents as

$$A_z = \frac{\mu I \Delta l}{4\pi r} \left[e^{-jk(r - h\cos\theta)} + e^{-jk(r + h\cos\theta)} \right]. \tag{8.93}$$

5. Figure 8.9 shows an array of N number of Hertzian dipoles whose current
   density is

Fig. 8.10 An infinitely long
line current

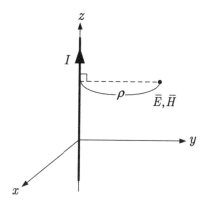

$$\overline{J}' = \hat{z} I \, \Delta l \, \delta(y') \, \delta(z') \sum_{n=0}^{N-1} \delta(x' - nd). \tag{8.94}$$

Evaluate the far field resulting from $\overline{J}'$.
Hint: The magnetic vector potential is

$$A_z = \mu I \, \Delta l \frac{e^{-jkr}}{4\pi r} \sum_{n=0}^{N-1} e^{jknd \sin\theta \cos\phi}. \tag{8.95}$$

6. Figure 8.10 shows an infinitely long line current I placed in free space. Assume that I is a constant invariant along z for simplicity. Evaluate the radiated magnetic field.
   Hint: Use the magnetic vector potential

$$\overline{A} = \frac{\mu}{4\pi} \int_{v'} \overline{J}' \frac{e^{-jk|\overline{r}-\overline{r}'|}}{|\overline{r}-\overline{r}'|} dv' \tag{8.96}$$

where $\overline{J}'$ is the current density. Note the integral formula

$$\int_{-\infty}^{\infty} \frac{e^{-jk\sqrt{\rho^2+u^2}}}{\sqrt{\rho^2+u^2}} du = -j\pi H_0^{(2)}(k\rho) \tag{8.97}$$

where $H_0^{(2)}(\cdot)$ is the Hankel function of the second kind of order zero.
7. The magnetic vector potential of a time-varying small current loop is given by (8.89), whereas the magnetic vector potential of a time-invariant small current loop is given by (3.69). Discuss the difference between (8.89) and (3.69).

Appendix A
Symbols, Notations, and Acronyms

A.1 Symbols

Symbol	Description
$\overline{A}, \overline{B}, \overline{C}, \cdots$	Vectors
$\hat{A}, \hat{B}, \hat{C}, \cdots$	Unit vectors
$\hat{x}, \hat{y}, \hat{z}$	Unit vectors in (x, y, z) coordinates
$\hat{\rho}, \hat{\phi}, \hat{z}$	Unit vectors in (ρ, ϕ, z) coordinates
$\hat{r}, \hat{\theta}, \hat{\phi}$	Unit vectors in (r, θ, ϕ) coordinates
$A, B, C, \cdots$	Scalars
$\overline{\mathcal{A}}, \overline{\mathcal{B}}, \overline{\mathcal{C}}, \cdots$	Time-varying quantities
$\overline{r} \, (= \hat{x}x + \hat{y}y + \hat{z}z)$	Position vector
$\hat{n}$	Normal unit vector
j	Imaginary unit where $j^2 = -1$

H. J. Eom, *Primary Theory of Electromagnetics*, Power Systems,
DOI: 10.1007/978-94-007-7143-7, © Springer Science+Business Media Dordrecht 2013

A.2 Notations and Units

Quantity	Notation	Unit	Abbreviation
Angular frequency	ω	radian/second	rad/sec
Capacitance	C	farad	F
Charge	Q	coulomb	C
Conductivity	σ	siemens/meter	S/m
Current	I	ampere	A
Current density	$\overline{J}$	ampere/meter^2	A/m^2
Dielectric constant	ϵ_r	–	–
Electric dipole moment	$\overline{p}$	coulomb-meter	C m
Electric displacement	$\overline{D}$	coulomb/meter^2	C/m^2
Electric field intensity	$\overline{E}$	volt/meter	V/m
Electric scalar potential	V	volt	V
Electric susceptibility	χ_e	–	–
Electromotive force	$\mathcal{V}$	volt	V
Energy	W	joule	J
Force	$\overline{F}$	newton	N
Frequency	f	hertz	Hz
Impedance	Z	ohm	Ω
Inductance	L	henry	H
Magnetic field intensity	$\overline{H}$	ampere/meter	A/m
Magnetic flux	Φ	weber	Wb
Magnetic flux density	$\overline{B}$	tesla	T
Magnetic susceptibility	χ_m	–	–
Magnetic vector potential	$\overline{A}$	weber/meter	Wb/m
Magnetization	$\overline{M}$	ampere/meter	A/m
Permeability	μ	henry/meter	H/m
Permittivity	ϵ	farad/meter	F/m
Polarization vector	$\overline{P}$	coulomb/meter^2	C/m^2
Power	P	watt	W
Poynting vector	$\overline{S}$	watt/meter^2	W/m^2
Resistance	R	ohm	Ω
Surface charge density	ρ_s	coulomb/meter^2	C/m^2
Surface current density	$\overline{J}_s$	ampere/meter	A/m
Voltage	V	volt	V
Volume charge density	ρ_v	coulomb/meter^3	C/m^3
Wavelength	λ	meter	m
Wave number	k	radian/meter	rad/m
Wave vector	$\overline{k}$	radian/meter	rad/m
Work	W	joule	J

A.3 Acronyms

Acronym	Description
AC	Alternating current
DC	Direct current
PEC	Perfect electric conductor
Re	Real part
TE	Transverse electric
TEM	Transverse electromagnetic
TM	Transverse magnetic
VSWR	Voltage standing-wave ratio

Appendix B
Vector Formulas

B.1 Additions and Products

$$\overline{A} \cdot \overline{A}^* = |\overline{A}|^2 \tag{B.1}$$

$$\overline{A} + \overline{B} = \overline{B} + \overline{A} \tag{B.2}$$

$$\overline{A} \cdot \overline{B} = \overline{B} \cdot \overline{A} \tag{B.3}$$

$$\overline{A} \times \overline{B} = -\overline{B} \times \overline{A} \tag{B.4}$$

$$(\overline{A} + \overline{B}) \cdot \overline{C} = \overline{A} \cdot \overline{C} + \overline{B} \cdot \overline{C} \tag{B.5}$$

$$(\overline{A} + \overline{B}) \times \overline{C} = \overline{A} \times \overline{C} + \overline{B} \times \overline{C} \tag{B.6}$$

$$\overline{A} \times (\overline{B} \times \overline{C}) = (\overline{A} \cdot \overline{C})\overline{B} - (\overline{A} \cdot \overline{B})\overline{C} \tag{B.7}$$

$$\overline{A} \cdot (\overline{B} \times \overline{C}) = \overline{B} \cdot (\overline{C} \times \overline{A}) = \overline{C} \cdot (\overline{A} \times \overline{B}) \tag{B.8}$$

B.2 Differential Operators

$$\nabla(a + b) = \nabla a + \nabla b \tag{B.9}$$

$$\nabla \cdot (\overline{A} + \overline{B}) = \nabla \cdot \overline{A} + \nabla \cdot \overline{B} \tag{B.10}$$

$$\nabla \times (\overline{A} + \overline{B}) = \nabla \times \overline{A} + \nabla \times \overline{B} \tag{B.11}$$

$$\nabla(ab) = a\nabla b + b\nabla a \tag{B.12}$$

$$\nabla \cdot (a\overline{A}) = a\nabla \cdot \overline{A} + \overline{A} \cdot \nabla a \tag{B.13}$$

$$\nabla \cdot \nabla a = \nabla^2 a \tag{B.14}$$

H. J. Eom, *Primary Theory of Electromagnetics*, Power Systems,
DOI: 10.1007/978-94-007-7143-7, © Springer Science+Business Media Dordrecht 2013

$$\nabla \times (a\overline{A}) = a(\nabla \times \overline{A}) - \overline{A} \times \nabla a \qquad (B.15)$$

$$\nabla \cdot (\overline{A} \times \overline{B}) = \overline{B} \cdot (\nabla \times \overline{A}) - \overline{A} \cdot (\nabla \times \overline{B}) \qquad (B.16)$$

$$\nabla \times \nabla \times \overline{A} = \nabla (\nabla \cdot \overline{A}) - \nabla^2 \overline{A} \qquad (B.17)$$

$$\nabla \cdot (\nabla \times \overline{A}) = 0 \qquad (B.18)$$

$$\nabla \times (\nabla a) = 0 \qquad (B.19)$$

Appendix C
Gradients, Divergences, Curls, and Laplacians

C.1 Rectangular Coordinates (x, y, z)

$$\nabla f = \hat{x}\frac{\partial f}{\partial x} + \hat{y}\frac{\partial f}{\partial y} + \hat{z}\frac{\partial f}{\partial z} \tag{C.1}$$

$$\nabla \cdot \overline{F} = \frac{\partial F_x}{\partial x} + \frac{\partial F_y}{\partial y} + \frac{\partial F_z}{\partial z} \tag{C.2}$$

$$\nabla \times \overline{F} = \hat{x}\left(\frac{\partial F_z}{\partial y} - \frac{\partial F_y}{\partial z}\right) + \hat{y}\left(\frac{\partial F_x}{\partial z} - \frac{\partial F_z}{\partial x}\right) + \hat{z}\left(\frac{\partial F_y}{\partial x} - \frac{\partial F_x}{\partial y}\right) \tag{C.3}$$

$$\nabla^2 f = \frac{\partial^2 f}{\partial x^2} + \frac{\partial^2 f}{\partial y^2} + \frac{\partial^2 f}{\partial z^2} \tag{C.4}$$

$$\nabla^2 \overline{F} = \hat{x}\nabla^2 F_x + \hat{y}\nabla^2 F_y + \hat{z}\nabla^2 F_z \tag{C.5}$$

C.2 Cylindrical Coordinates (ρ, ϕ, z)

$$\nabla f = \hat{\rho}\frac{\partial f}{\partial \rho} + \hat{\phi}\frac{1}{\rho}\frac{\partial f}{\partial \phi} + \hat{z}\frac{\partial f}{\partial z} \tag{C.6}$$

$$\nabla \cdot \overline{F} = \frac{1}{\rho}\frac{\partial(\rho F_\rho)}{\partial \rho} + \frac{1}{\rho}\frac{\partial F_\phi}{\partial \phi} + \frac{\partial F_z}{\partial z} \tag{C.7}$$

$$\nabla \times \overline{F} = \hat{\rho}\left(\frac{1}{\rho}\frac{\partial F_z}{\partial \phi} - \frac{\partial F_\phi}{\partial z}\right) + \hat{\phi}\left(\frac{\partial F_\rho}{\partial z} - \frac{\partial F_z}{\partial \rho}\right)$$
$$+ \hat{z}\left[\frac{1}{\rho}\frac{\partial(\rho F_\phi)}{\partial \rho} - \frac{1}{\rho}\frac{\partial F_\rho}{\partial \phi}\right] \tag{C.8}$$

H. J. Eom, *Primary Theory of Electromagnetics*, Power Systems,
DOI: 10.1007/978-94-007-7143-7, © Springer Science+Business Media Dordrecht 2013

$$\nabla^2 f = \frac{1}{\rho}\frac{\partial}{\partial\rho}\left(\rho\frac{\partial f}{\partial\rho}\right) + \frac{1}{\rho^2}\frac{\partial^2 f}{\partial\phi^2} + \frac{\partial^2 f}{\partial z^2} \tag{C.9}$$

$$\nabla^2 \overline{F} = \nabla\left(\nabla\cdot\overline{F}\right) - \nabla\times\nabla\times\overline{F} \tag{C.10}$$

C.3 Spherical Coordinates (r, θ, ϕ)

$$\nabla f = \hat{r}\frac{\partial f}{\partial r} + \hat{\theta}\frac{1}{r}\frac{\partial f}{\partial\theta} + \hat{\phi}\frac{1}{r\sin\theta}\frac{\partial f}{\partial\phi} \tag{C.11}$$

$$\nabla\cdot\overline{F} = \frac{1}{r^2}\frac{\partial(r^2 F_r)}{\partial r} + \frac{1}{r\sin\theta}\frac{\partial(\sin\theta F_\theta)}{\partial\theta} + \frac{1}{r\sin\theta}\frac{\partial F_\phi}{\partial\phi} \tag{C.12}$$

$$\nabla\times\overline{F} = \frac{\hat{r}}{r\sin\theta}\left[\frac{\partial(F_\phi\sin\theta)}{\partial\theta} - \frac{\partial F_\theta}{\partial\phi}\right] + \frac{\hat{\theta}}{r}\left[\frac{1}{\sin\theta}\frac{\partial F_r}{\partial\phi} - \frac{\partial(r F_\phi)}{\partial r}\right]$$

$$+ \frac{\hat{\phi}}{r}\left[\frac{\partial(r F_\theta)}{\partial r} - \frac{\partial F_r}{\partial\theta}\right] \tag{C.13}$$

$$\nabla^2 f = \frac{1}{r^2}\frac{\partial}{\partial r}\left(r^2\frac{\partial f}{\partial r}\right) + \frac{1}{r^2\sin\theta}\frac{\partial}{\partial\theta}\left(\sin\theta\frac{\partial f}{\partial\theta}\right)$$

$$+ \frac{1}{r^2\sin^2\theta}\frac{\partial^2 f}{\partial\phi^2} \tag{C.14}$$

$$\nabla^2 \overline{F} = \nabla\left(\nabla\cdot\overline{F}\right) - \nabla\times\nabla\times\overline{F} \tag{C.15}$$

Appendix D
Dirac Delta Functions

The Dirac delta function $\delta(\bar{r} - \bar{r}')$ is defined as

$$\delta(\bar{r} - \bar{r}') = 0 \quad \text{when } \bar{r} \neq \bar{r}' \tag{D.1}$$

$$\int_v \delta(\bar{r} - \bar{r}')dv = 1 \quad \text{when } v \text{ contains } \bar{r}' \tag{D.2}$$

where $\bar{r}$ and $\bar{r}'$ are the position vectors as

$$\bar{r} = \hat{x}x + \hat{y}y + \hat{z}z \tag{D.3}$$

$$\bar{r}' = \hat{x}x' + \hat{y}y' + \hat{z}z'. \tag{D.4}$$

For instance, the Dirac delta function in three-dimensional rectangular coordinates is written as

$$\delta(\bar{r} - \bar{r}') = \delta(x - x')\delta(y - y')\delta(z - z') \tag{D.5}$$

where

$$\delta(x - x') = 0 \quad \text{when } x \neq x' \tag{D.6}$$

$$\int_{-\infty}^{\infty} \delta(x - x')dx = 1. \tag{D.7}$$

The Dirac delta function has the sifting property

$$\int_v f(\bar{r})\delta(\bar{r} - \bar{r}')dv = f(\bar{r}') \quad \text{when } v \text{ contains } \bar{r}'. \tag{D.8}$$

H. J. Eom, *Primary Theory of Electromagnetics*, Power Systems,
DOI: 10.1007/978-94-007-7143-7, © Springer Science+Business Media Dordrecht 2013

Appendix E
Answers to Problems

Chapter 1

1. $\hat{x} = \hat{r}\sin\theta\cos\phi + \hat{\theta}\cos\theta\cos\phi - \hat{\phi}\sin\phi$.

3. $\nabla f = \hat{\rho}\dfrac{\partial f}{\partial\rho} + \hat{\phi}\dfrac{1}{\rho}\dfrac{\partial f}{\partial\phi} + \hat{z}\dfrac{\partial f}{\partial z}$.

4. $\nabla^2 f = \dfrac{1}{r^2}\dfrac{\partial}{\partial r}\left(r^2\dfrac{\partial f}{\partial r}\right) + \dfrac{1}{r^2\sin\theta}\dfrac{\partial}{\partial\theta}\left(\sin\theta\dfrac{\partial f}{\partial\theta}\right) + \dfrac{1}{r^2\sin^2\theta}\dfrac{\partial^2 f}{\partial\phi^2}$.

9. 0 .

10. $-\dfrac{\hat{x}(x-x') + \hat{y}(y-y') + \hat{z}(z-z')}{[(x-x')^2 + (y-y')^2 + (z-z')^2]^{1.5}}\left(= -\dfrac{\hat{R}}{R^2}\right)$.

Chapter 2

1. $E_z = \begin{cases} \dfrac{\rho_s}{2\epsilon_0}\left[1 - \dfrac{z}{(z^2+a^2)^{1/2}}\right] & \text{for } z > 0 \\[4mm] -\dfrac{\rho_s}{2\epsilon_0}\left[1 + \dfrac{z}{(z^2+a^2)^{1/2}}\right] & \text{for } z < 0. \end{cases}$

2. $E_\rho = \begin{cases} \dfrac{Q}{2\pi\epsilon_0\rho} & \text{for } \rho \geq a \\[4mm] \dfrac{Q\rho}{2\pi\epsilon_0 a^2} & \text{for } \rho \leq a. \end{cases}$

3. $E_x = \begin{cases} 0 & \text{for } x \leq -a \\[2mm] \dfrac{\rho_v}{\epsilon_0}(x+a) & \text{for } -a \leq x \leq 0 \\[2mm] \dfrac{\rho_v}{\epsilon_0}(-x+a) & \text{for } 0 \leq x \leq a \\[2mm] 0 & \text{for } a \leq x. \end{cases}$

The field is shown in Fig.E.1.

4. $\overline{E}_2\big|_{y=0} = \hat{x}E_x + \hat{y}\dfrac{\epsilon_1}{\epsilon_2}E_y$.

5. $C = \dfrac{\pi\epsilon_0}{\ln(d/a)}$.

H. J. Eom, *Primary Theory of Electromagnetics*, Power Systems,
DOI: 10.1007/978-94-007-7143-7, © Springer Science+Business Media Dordrecht 2013

Fig. E.1 Electric field E_x due to charged slabs where $E_0 = \rho_v a / \epsilon_0$

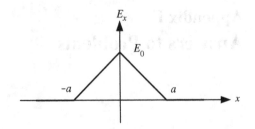

6. $W_e = \dfrac{4\pi \rho_v^2 a^5}{15\epsilon_0}$.

7. $V = \begin{cases} \dfrac{Q}{4\pi\epsilon_0 r} & \text{for } r \geq a \\[3mm] \dfrac{Q}{4\pi\epsilon_0 a} & \text{for } r \leq a. \end{cases}$

8. $W_e = \dfrac{Q^2}{4\pi\epsilon l} \ln(b/a)$.

9. $V = \displaystyle\sum_{n=1,3,5,\cdots}^{\infty} \dfrac{4}{n\pi} \dfrac{\left\{ V_1 \sinh\left(\dfrac{n\pi}{b}x\right) + V_0 \sinh\left[\dfrac{n\pi}{b}(a-x)\right] \right\}}{\sinh\left(\dfrac{n\pi}{b}a\right)} \sin\left(\dfrac{n\pi}{b}y\right)$.

10. $E_\rho = \dfrac{\rho l}{2\pi\epsilon_0 \rho}$.

Chapter 3

1. $R = \dfrac{1}{2\pi\sigma l} \ln \dfrac{b}{a}$.

2. $H_z = \dfrac{I}{2a}$.

3. $B_z = \dfrac{\mu_0 \rho_s \omega}{2} \left(\dfrac{a^2 + 2z^2}{\sqrt{a^2 + z^2}} - 2z \right)$.

4. $H_x = \begin{cases} J_s/2 & \text{for } z > 0 \\ -J_s/2 & \text{for } z < 0. \end{cases}$

5. Force per unit length $= \hat{z}\dfrac{\mu_0 I^2}{\pi w} \tan^{-1}\left(\dfrac{w}{2h}\right)$.

6. $H_\phi = \dfrac{I}{2\pi\rho}$.

Chapter 4

3. $\mathcal{V} = \omega Bab \sin \omega t$.

4. $L = \dfrac{\mu_0}{\pi} \ln \dfrac{d}{a}$.

5. $L_{12} = \dfrac{\pi\mu_0 a^2 b^2}{2h^3}$.

Chapter 5

1. The boundary conditions are

$$-V_1 + V_2 = 0$$

$$-\epsilon_1 \frac{\partial V}{\partial n}\Big|_{V=V_1} + \epsilon_2 \frac{\partial V}{\partial n}\Big|_{V=V_2} = \rho_s.$$

Chapter 6

4. $\theta_i = \tan^{-1}\sqrt{\epsilon_2/\epsilon_1}$.
5. If the incident angle θ_i is greater than $\theta_c = \sin^{-1}\sqrt{\epsilon_2/\epsilon_1}$, total reflection occurs.
6. The reflected wave is $\overline{E} = -(\hat{x} + j\hat{y})e^{-jkz}$, which is left-hand circularly polarized.
7. Reflected power $= 0$.

Chapter 7

1. $P_{av} = \pi\sqrt{\dfrac{\epsilon}{\mu}}\dfrac{|V|^2}{\ln(b/a)}$.

2. $P_{av} = \dfrac{|V_s|^2}{8Z_0}\left(1 - \left|\dfrac{Z_l - Z_0}{Z_l + Z_0}\right|^2\right)$.

3. $\Gamma = \dfrac{Z_l - Z_1}{Z_l + Z_1}$.

4. $Z = \infty$.

5. The voltage sketch is shown in Fig. E.2.

Fig. E.2 A response to the rectangular pulse

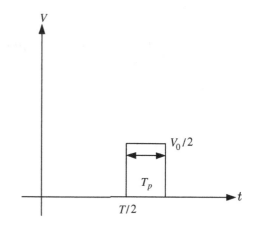

Chapter 8

1. 6.52 GHz $<$ frequency < 13.04 GHz .

2. $P_{av} = |E_0|^2 \dfrac{ab}{8} \dfrac{k_z \omega \epsilon}{\left(k_x^2 + k_y^2\right)}$ where $k_x = m\pi/a$, $k_y = n\pi/b$, and $k_z =$

$\sqrt{k^2 - k_x^2 - k_y^2}$.

3. The TM wave propagates when $k > m\pi/a$ as

$$E_z = E_0 \sin k_x x e^{-jk_z z}$$

$$E_x = \frac{-jk_z}{k_x} E_0 \cos k_x x e^{-jk_z z}$$

$$H_y = \frac{-j\omega\epsilon}{k_x} E_0 \cos k_x x e^{-jk_z z}$$

where $k_x = m\pi/a$, $k_z = \sqrt{k^2 - (m\pi/a)^2}$, and m is an integer ($m \geq 1$).

4.

$$E_\theta = j\omega \frac{\mu e^{-jkr}}{2\pi r} I \Delta l \sin\theta \cos(kh\cos\theta)$$

$$H_\phi = E_\theta / \eta.$$

5.

$$E_\theta = j\omega\mu I \Delta l \sin\theta \frac{e^{-jkr}}{4\pi r} \exp\left[\frac{jk(N-1)d}{2} \sin\theta\cos\phi\right] \frac{\sin(\frac{kNd}{2}\sin\theta\cos\phi)}{\sin(\frac{kd}{2}\sin\theta\cos\phi)}$$

$$H_\phi = E_\theta / \eta.$$

6. $H_\phi = -\dfrac{jkI}{4} H_1^{(2)}(k\rho)$, where $H_1^{(2)}(k\rho)$ is the Hankel function of the second kind of order one.

Index

H. J. Eom, *Primary Theory of Electromagnetics*, Power Systems
DOI: 10.1007/978-94-007-7143-7, © Springer Science+Business Media Dordrecht 2013

Printed in the United States
By Bookmasters